Making Health Work

Studies in Demography
General Editors
Eugene A. Hammel
Ronald D. Lee
Kenneth W. Wachter

Making Health Work

Human Growth in Modern Japan

Carl Mosk

UNIVERSITY OF CALIFORNIA PRESS
Berkeley · *Los Angeles* · *London*

University of California Press
Berkeley and Los Angeles, California

University of California Press
London, England

Copyright © 1996 by The Regents of the University
of California

Library of Congress Cataloging-in-Publication Data

Mosk, Carl.
    Making health work : human growth in modern Japan / Carl Mosk.
        p. cm.—(Studies in demography : 8)
    Includes bibliographical references and index.   ·
    ISBN 0-520-08315-6 (alk. paper)
        1. Quality of life—Japan.   2. Human growth—Japan.
    3. Anthropometry—Japan.   4. Japan—Population.   I. Title.
    II. Series: Studies in demography (Berkeley, Calif.) : 8.
    HN724.M68   1996
    573'.6'0952—dc20   96-26629
                                                                 CIP

Printed in the United States of America

1   2   3   4   5   6   7   8   9

*to my mother, Mary B. Hanley
and to the memory of my father,
Sanford A. Mosk*

Contents

Illustrations

CHARTS

FIGURE

MAPS

Preface and
Acknowledgments

The subject of this study is population quality in Japan. It is my intention to make a strong case for population quality as a useful variable and one that can be effectively measured by height and weight and chest girth and body mass index for schoolchildren. I also hope to convince the reader that the secular improvement in population evident for Japan over the last several centuries is a result of improvements in net nutritional intake, that is, improvements in gross nutritional intake net of the nutrients burned up in combating disease and in fueling physical work effort. Finally, I hope to firmly establish that while technology and the shift away from agriculture and toward manufacturing matters a great deal to the improvement in net nutritional intake and hence to the enhancement of population quality, demand, as enunciated through the vehicle of political protest movements, also matters as well. Hence I ask the reader of this book to wear two hats, that of the statistician and that of the social historian.

This work arises from my stubborn conviction that enhancing population quality is a key ingredient to successful economic growth in our era. Telling the story for Japan, a country that perhaps better than any other exemplifies the capacity to rapidly industrialize and raise levels of income per capita despite severe natural resource constraints, will, I trust, make convincing the critical importance of population quality.

During the course of the research that underlies this account I have acquired numerous debts and obligations, too many to adequately acknowledge here. However, I would like to single out my debts to three Japanese scholars who have greatly assisted me with my research on Japan's population over the last five years: Hiroshi Kawaguchi of Tezukayama University, Osamu Saito of Hitotsubashi University, and Yasukichi Yasuba of Osaka Gakuin University. In addition, I want to thank Isao Ohashi of Nagoya University for inviting me to spend three months in Nagoya during the spring and summer of 1994 as a visiting scholar and to the staff of the Economic Research Center at Nagoya University for assisting me in my research there. As a result of their invaluable assistance I was able to assemble a large amount of data on anthropometric measures for farmers and factory workers, some of which I discuss here. Seminar participants at talks I gave at Osaka Gakuin University, Ritsumeikan University, and Tezukayama University helped me clarify my thinking on a variety of issues concerning Japanese demographic history, and I am extremely thankful for the comments made by these individuals. My friend Yoshi-fumi Nakata of Doshisha University assisted me in arranging my trips to and within Japan and I am in debt to him for his efforts on my behalf.

At the University of Victoria I have benefited from the comments and criticisms of many colleagues, some made in seminars and some made in private conversation. I would like to offer special thanks here to Kenneth Avio, Judith Giles, David Giles, and Malcolm Rutherford in the Department of Economics; to Daniel Bryant in the Department of Pacific and Asian Studies; and to Eric Roth in the Department of Anthropology. They have contributed to this volume, although in ways they may find surprising. And I wish to acknowledge the assistance of the very professional staff of the Department of Economics at the University of Victoria, especially Pattie Eccleston, Barbara Provan, and Priscilla Shiu. I also wish to express my thanks to Ole Heggen of the Geography Department staff for preparing the three maps and the figure. Finally I want to acknowledge the helpful comments of participants and panelists at a session of the 1994 annual meeting of the Social Science History Association which concerned historical trends in height and at which I presented an early version of some of the arguments advanced in chapter 2 of this work.

By way of closing out my acknowledgment of academic assistance, I would like to mention the great support and assistance of Eugene Hammel of the University of California, Berkeley, who has aided me in countless ways over the course of my academic life, with earlier projects as well as with this one.

The writing of a book places a strain on those one loves. I want to thank Kumiko for bearing with me when I was deeply involved in writing this manuscript and for encouraging me to "put on my thinking cap" and to "get to the point." I hope that she feels I have done so in this book. If she does, it will be my greatest reward.

This book is dedicated to my mother, Mary B. Hanley, and to the memory of my father, Sanford A. Mosk, who was a true gentleman and a scholar. He always emphasized the fact that a book must have a central thesis. I believe this work does have a central thesis and therefore fits his definition of a real book. But that is for the reader to judge.

Victoria, British Columbia

PART I

Secular Trend, 1900–1985

CHAPTER I

Introduction

"There were giants in the earth in those days," says Genesis, drawing a powerful analogy between secular diminution in physique and degradation in moral fiber.[1] Linking dramatic change in quality to decisive changes in physique as it does, the quotation is an apt starting point for this work. My two key theses—that improvements in net nutritional intake mainly attributable to a diminution in the demands placed upon it caused by a shift out of labor-intensive work and a decline in the frequency and duration of infection due to more effective medical and public health intervention led to the improvement in population quality interpreted in terms of enhanced capability and work capacity; and that in addition to supply factors like technology, demand factors working both through the market and through entitlements were crucial to this secular trend—are theses about a striking secular change that has rendered Japan's young adult population potentially more productive on a per capita basis and physically taller and heavier than its forbears. Exaggerating somewhat for the sake of emphasis we can say that in quality terms, Japan's population today, in comparison to her population in the past, consists of giants.

Just as the quote from Genesis argues by way of analogy and metaphor, so does the account offered here. But while analogy is important in this work, so is the marshaling and interpretation of historical evidence. In this study I scrutinize a considerable body of evidence, in an eclectic manner. My account is eclectic because it combines both qualitative and quantitative data and because insofar as it has a quantitative focus it draws on a wide variety of techniques to process and present that data—for instance, regression analysis conventionally employed by economists and cross-classification of data favored by sociologists and anthropologists—in weaving an account that I hope will be comprehensible to historians as well as specialists in various social sciences. That I have adopted an eclectic approach reflects not only my own predilections in analytical matters (in particular, my belief that it is through an interaction of inductive and deductive techniques that progress is secured) but also my desire to communicate my findings to scholars in a wide range of disciplines. I believe this is important because the secular improvement in population quality in Japan was a broad social process going beyond the mere operation of markets as they evolve in response to changes in technology. Institutions at the local and national levels governing the distribution of entitlements over foodstuffs and health-enhancing public health and medical services and the demand for these institutions as voiced through political and social protest play a very important role in my account.

Of the analogies and metaphors central to this volume, the most important are those involving the definition and measurement of population quality and nutritional intake net of the demands placed on it. The remainder of this chapter briefly discusses these key metaphors and some of the thorniest conceptual issues surrounding them and clears the way for more detailed treatment in terms of the historical data for Japan that I analyze in the remainder of this book. I begin with the concept of population quality.

POPULATION QUALITY
AND THE STANDARD OF LIVING

By the average quality of a population, I mean the average potential capability and potential work capacity of its members. Capability and work capacity are to be understood in both physical and mental terms. Whether a population of a given average capacity or capability chooses to exercise and develop the potential it has depends on the attitudes toward and the incentives that individuals face concerning the development of these capabilities. From a theoretical viewpoint my concept of average population quality is quite close to the notion put forward by A. Sen et al. (1987) concerning the proper definition of the standard of living. The similarity of my view is briefly discussed in the remainder of this section. From a practical empirical viewpoint, I choose to use anthropometric measures concerning height, weight, and chest girth for children and young adults at various key ages. That these measures serve as useful proxies for potential capabilities and work capacity is a matter that, to the best of my knowledge, is not discussed by Sen. Discussion of the usefulness of employing these measures as proxies is taken up below.[2]

Why should we focus on population quality as defined in terms of capabilities and work capacity? There are two lines of reasoning that lead us to this conclusion. The first is the potential feedback of labor productivity on future productivity. That there is a relationship between the productivity of the past and the productivity of the present arises from a variety of potential linkages. For instance, short-run linkages run through the level of nutritional intake or a sense of well-being to work capacity or the eagerness to work in the present period or in the immediate future. Or again this process may take place over several generations. The better fed and the healthier women are in any given period, the more likely they are to give birth to offspring with adequate birth weights and the more nutritious their breast milk is if they elect to breast feed their infants. That feedback appears to have been empirically important in the Japanese case is crucial to the argument I develop in Part II of this book concerning the motivation for and role of population quality-enhancing behavior on the part of households and non-household organizations like enterprises and governments.

The second reason to start with a definition centered on capabilities and work capacity is because it leads to a linkage between the measurement of population quality proposed in this volume and the measurement of the standard of living as proposed by Sen (1987). In my opinion Sen's definition has much to offer when it is contrasted with the more conventional definitions used by economists. For instance, Sen (1987: 7 ff.) forcibly argues that the standard subjectivist utilitarian view runs into the objection not only that utility in and of itself cannot be directly measured but that definitions employing it fall short of what most people believe on logical grounds is the standard of living. Thus if we interpret utility in terms of pleasure and happiness, we must address the question of whether persons who are very poor and exploited in material terms may not, by dint of being socially conditioned into lacking ambition and material aspirations, find happiness in their poverty equal to or greater than that experienced by a wealthy individual who commands many more resources in consumption. Again if we think of utility as defined in terms of the satisfaction of desire, we encounter the same objection raised immediately above that the downtrodden may find their desires so stifled through progressive disappointments that they abandon these desires one by one, retaining only a few modest ones that they are sure of fulfilling. Despite the fact that such a downtrodden individual may well be satisfying the few desires he or she is left with, is it not unreasonable to say that such an individual is well-off? Again if we conceive of utility in terms of choice, we encounter a host of problems: what about choices made by an individual which benefit others and not the individual per se? What about choices involving nonmarketed goods like one's own children? Again if two individuals with identical time and income constraints choose different bundles of commodities, can we conclude anything about relative levels of standard of living?

It is theoretically possible to avoid the pitfalls of subjectivist definitions by using consumption, or what Sen (1987: 14 ff.) calls opulence. Now ignoring the question of constructing price indexes so that comparisons between two different economies or within one economy over time can be made, which plagues the actual empirical estimates of relative opulence that are made, there remain conceptual problems with using consumption of goods and services for individuals in the same economy facing identical prices. The two individuals may have different physical needs, for example, different demands for foodstuffs because one individual is ill and the other is not, or because one individual has a higher metabolic rate than the other, or because one lives in a colder climate than the other, and so forth. In practice there is no way to control for all background variables that condition demands placed on consumption so that unambiguous quantitative rankings of individuals can be arrived at. While it is reasonable to suppose that a doubling of income per capita is associated with some improvement in the standard of living, if this gain in income is at least partly purchased at the expense of increased crime or environmental degradation or an increase in stress, it is not at all clear to what extent the overall standard of living has actually improved.

Of course, there are many thoughtful advocates of the conventional approaches using either opulence or opulence taken together with other indicators of quality of life, the nonmaterial components of welfare implicitly being purchased from the resources made available by opulence. For the sake of my ongoing comparison between the measures of population quality that I espouse and the various facets of opulence, specifically, net nutritional intake, which I will make use of in my empirical analysis as causal factors explaining how levels of population quality are determined, I present some indicators of opulence and associated quality of life measures for Japan over the period 1881–1980 in table 1. Note that while death rates go down and life expectancies go up roughly in tandem with the rise in per capita income and consumption, reported illness rates for middle-aged males also go up. And note that the trend in reported illness has a direct bearing on some of Sen's objections to the conventional subjectivist/utilitarian approaches to measuring the standard of living. Is the trend due to increases in levels of income that allow people to "purchase" more leisure by declaring themselves ill in circumstances that in the past, when income was far lower, would not have served as socially acceptable justification for taking time off? Or are middle-aged men in Japan subject to more stress now than in the past because expectations about work capacity have risen faster than work capacity? Or is the trend a mirror for improved quality of medical care and diagnosis?

In any case, to return to the conceptual issue raised by Sen, he rejects both subjectivistic utilitarianism and opulence as criteria for defining the standard of living. As an alternative he proposes a definition that includes the various "doings" a person achieves; that is, he advocates use of a definition based on achievement or capability (the capacity to achieve). It is not necessary here to go into detail concerning the very interesting distinctions Sen makes between functionings that are achievements and capabilities that are the capacities to achieve and are conditioned, among other things, by freedom. What is relevant to our discussion here is that Sen focuses on what I think many, perhaps most, people mean when they talk about a life "well lived," or "richly lived." People have in mind some concept of the capacity to achieve things, whether it is because they are physically capable of doing things or because they are skilled or well educated or sensitive to opportunities, and so forth. Without taking up the thorny issue of political constraints on freedom to exert oneself, I take a narrow form of this concept of achievement and capacity to achieve and define it as I have done above in terms of population quality. Certainly there is a relationship between population quality as I define it and Sen's concept of the standard of living, but because my definition is narrower than his, populations of identical quality according to my definition may well enjoy quite different levels of the standard of living, say, because of differences in the capital-to-labor ratio or in the political system. For the purposes of this

TABLE 1

Per Capita Income and Consumption and Indicators of Health in Japan: Selected Years, 1881-1980

| | Per Capita Income and Consumption[a] | | | | | Indicators of Health[b] | | | | | |
| | | Consumption | | | | Life Expectancy, Age 0 | | Deaths per 100,000 Persons | | Illness Rates per 100,000 Males | |
Years	GDPPC	Total	Food	CAL	CALW	Males	Females	TB	Pne/Bro	35-44	45-54
1881-1885	103.2	92.0	60.3	1705	2252	n.e.	n.e.	n.e.	n.e.	n.e.	n.e.
1901-1905	140.6	122.3	73.6	2143	2838	n.e.	n.e.	179.2	226.6	n.e.	n.e.
1921-1925	208.5	174.1	100.5	2423	3208	42.1	43.2	212.6	305.0	n.e.	n.e.
1931-1935	225.7	181.9	95.5	2309	3048	46.9	49.6	186.3	214.4	n.e.	n.e.
1951-1955	276.7	166.6	81.4	2096	n.e.	50.1	54.0	93.6	73.7	4,550	6,130
1966-1970	931.8	503.0	340.1	2219	n.e.	69.3	74.7	18.8	31.5	8,620	12,660
1976-1980	n.e.	n.e.	n.e.	n.e.	n.e.	73.3	78.8	7.8	30.7	7,430	12,180

SOURCES: Various tables from Japan Office of the Prime Minister (various years); Japan Statistical Association 1987, 1988; Mosk and Pak 1978; and Ohkawa and Shinohara 1979.

NOTES: [a]GDPPC = Gross domestic product per capita in 1934-1936 prices (total and food consumption also in 1934-1936 prices); CAL = calories consumed per day per capita; CALW = calories consumed per day per consumer unit weighted population, where weights are as follows: (1) for males in the age groups 0-4, .4413; 5-9, .7100; 10-14, .9; 15-19, 1.0167; 20-39, 1.0; 40-49, .95; 50-59, .95; 60 and over, .75. (2) for females in the age groups 0-4, .4367; 5-9, .6667; 10-14, .8; 15-19, .7833; 20-39, .7333; 40-49, .6967; 50-59, .66; 60 and over, .55.

[b]Life expectancy figure for 1931-1935 is for 1935-1936; figure for 1951-1955 is for 1947; figure for 1956-1960 is for 1960. TB = tuberculosis; Pne/Bro = pneumonia and bronchitis (TB and Pne/Bro figures are for 1900-1904 rather than 1901-1905, etc.). Illness figures were taken during a three-day period in the fall of each year. Illness figure for 1951-1955 is for 1955; figure for 1966-1970 is for 1970; figure for 1976-1980 is for 1980.

n.e. = not entered (or available).

study a narrow definition will suffice. Moreover, to actually *measure* the standard of living it is necessary to narrow the scope of our definition even further. Which is the issue to which we will now proceed.

ANTHROPOMETRIC MEASURES
AND THE SECULAR TREND IN HUMAN GROWTH

I propose to measure population quality in terms of anthropometric measures for children and young adults, namely, in terms of average levels of height, weight, chest girth, and weight for height indexes, for males and females at various ages up to twenty. To understand why this approach can usefully serve to measure quality, it is necessary to briefly consider three major findings from the field of auxology on which I draw in selecting my measures. These three findings are as follows. (1) There has been a secular trend in the anthropometric measures for adults in all populations that have experienced a pronounced and sustained rise in per capita income, referred to hereafter as the secular trend in levels of human growth. (2) There has also been a secular trend in the tempo or timing of growth in children and youths as they mature toward their terminal adult heights and weights and chest girth. The mean age of maturation has declined; in particular, the mean age when children experience their greatest postinfancy growth has declined over time. This is referred to hereafter as the secular trend in the tempo of human growth. (3) While the gene pool and heredity are important at the individual level and at the average level over a period of many generations, evolution within the gene pool is a minor factor in the secular trend. I take up the third item in the next section and turn to the first two points in the remainder of this section.

Auxology is a field with a venerable tradition reaching back to the Renaissance. J. M. Tanner (1981), one of the leading authorities in the field, has provided us with an engaging account of the evolution of the study of human dimensions from its beginnings in the work of artists like Leonardo da Vinci and Albrecht Dürer to the studies by Montbeillard of his son's growth in the mid-eighteenth century to the anthropological work of Shuttleworth and Franz Boas. Among other findings to emerge from this literature is an appreciation that the process of human growth is quite uniform for each sex taken separately as evidenced by a regular age-specific profile for the growth process and by the simultaneous coordination or coincidence of organ and tissue development for the various components of the body, but that development age and chronolog-

ical age vary individually within populations and over time, in a secular
sense, for populations. For instance, Tanner (1961) shows that the development of the brain and the ability to perform well on intelligence
tests is related to physical maturation in other areas of the body, like the
length of legs and arms and the size and functioning of the sexual organs. Now because of the long history during which auxologists have
developed a set of precise measurements for exact calculation of height,
weight, and other physical characteristics, and because auxological data
is often generated by military and educational institutions as by-prod-
ucts of their examination of physical fitness for the individuals under
their command or charge, we can document the secular trend in anthropometric measures for a number of national populations or subpopulations.

And wherever we can document these trends, it appears that they ac-
company economic and social modernization (see Eveleth and Tanner
1990; Tanner 1978, 1994; and the various chapters and preface by
Komlos in Komlos 1994). Hence researchers have been given the opportunity to document the secular improvement of living standards by
using data on secular movements in anthropometric measures. And this
brings us back full circle to the issue of the standard of living and to my
position regarding how to interpret secular changes in the level and
tempo of human growth.

Note that first of all I do not define or measure the standard of living
in terms of the anthropometric measures; rather, I define population
quality in terms of capabilities and work capacity, which I measure in
terms of the anthropometric measures for children and young adults. By
measuring population quality in terms of the properties of children and
young persons, many of whom have not yet entered into employment, I
explicitly focus on capabilities and capacity for future work as opposed
to achievements and accomplishments. Thus population quality in my
definition is very close to a narrow version of Sen's concept of the standard of living, which excludes the political and social constraints that
may in practice limit the ability of individuals to turn capacities into realized achievements and concentrates on potential adult physical and
mental work capacity. Other things being equal, the greater the population quality, the greater the standard of living in Sen's sense. The relationship between population quality and the standard of living defined
in other terms—say, in subjective utilitarian terms or in terms of opulence—is complex and far less clear than is the relationship between
population quality and Sen's standard of living concept. Suffice it to say

that there is no obvious connection between my concept of population quality and the utilitarian definitions, but that some aspects of the standard of living defined in terms of opulence do play a role in my account as determinants of population quality. It is important to keep in mind that when I refer to the standard of living, I am referring to Sen's concept, and when I refer to the determinants of population quality, I have in mind specific variables that are often incorporated into the opulence definition of the standard of living.

COEVOLUTION

I hope it is by now clear that I do not subscribe to the view that the standard of living, particularly defined in terms of opulence, is equivalent to population quality. While I have already provided a number of grounds for reaching this conclusion, I have not yet considered one of the most compelling, namely, the influence of the gene pool on anthropometric measures. R. Steckel (1994b: 1, 9 ff.) argues that as long as we work with population averages and changes over time or differences between population averages, we can largely control for the influence of the gene pool. He does concede, however, that comparisons between populations of Asian and Western European descent are complicated by genetic factors. For instance, P. Eveleth and J. M. Tanner (1990: chap. 9) provide an abundance of evidence that physical proportions—for example, leg to trunk length as measured by the ratio of sitting to standing height—vary between different gene pools: Individuals of African descent tend to have long legs relative to trunk length; individuals of European descent tend to have moderate leg lengths relative to trunk length; and individuals of Asian descent tend to have short legs relative to trunk length. For this reason, other things being equal, adults of African descent tend to be taller than individuals of Asian descent. That there are observed height differentials does not necessarily speak to the question of whether long-standing or contemporary opulence-based standard of living differentials exist among these groups.[3] Now it may be thought that while international comparisons are complicated by genetic factors, comparisons within a gene pool or secular changes within a gene pool are free of this problem. Is it not the case, for example, that secular change for, or differentials within, Japan's population (which is often said to be racially homogeneous due to its isolation from the Eurasian landmass) are unambiguously attributable to factors that are not genetic? Unfortunately, as we shall now see, this position cannot be sustained.

First, race itself is a questionable category in anthropological analysis. The prevailing view is that while genetic inheritance is important,
there is no such thing as distinct "races." Within the boundaries of
Japan live persons of Ainu, Korean, and Chinese descent and/or prog-
eny descended from marriages between persons of different ethnic origin.[4] For this reason, throughout this book I refer to "Japan's population" rather than to "the Japanese people."

Second, coevolution may exist. By coevolution, I mean the interact-
ing evolution of culture with genes. Imagine that there is random and
ongoing genetic change and that most of these genetic changes vanish
over time but that some have adaptive or survival value because of the
cultural environment in which the phenotypes carrying the genetic cod-
ing exist.[5] The possibility of coevolutionary change has been extensively
explored by anthropologists in the last several decades. A number of the
arguments in this field are systematically reviewed and tested by W.
Durham (1991). Several examples culled from the literature appear in
chart 1, below. Note that two major arguments, both of a coevolutionary nature, have been advanced to explain why persons of Asian descent tend to have shorter legs than Africans. One is that because of random genetic changes selected for, because marriage ages were unusually
youthful in Asia, or because of diet, Asians go through the adolescent
growth spurt approximately a year earlier than non-Asians. Hence, because in the years leading up to the adolescent growth spurt legs are favored in growth, non-Asians have an extra year or more for leg development. The second main coevolutionary argument relates to climatic
differences: gene pools that evolved in hot climates require greater heat
loss per unit of volume; hence genetic changes that produce longer legs
are favored. Insofar as coevolution does occur, does it not occur very
slowly, over generations and hence over hundreds or even thousands of
years? In the short span of a century or perhaps two centuries is it not
reasonable to suppose that there is too little time for coevolution to
occur? For instance, is it not reasonable to suppose that the secular gain
in standing height that I will document for Japan's population over the
period 1900–1985 is due to nongenetic factors?

At first glance the argument appears plausible, but there are problems
with it. Taken literally, this thesis means that the typical male living in
Japan at the turn of the century had the genetic potential to reach an adult
height of, say, on average 170 centimeters but, due to the exertions of
physical work, the ravages of infection, and inadequate nutrition, was
only able to reach a level of around 160 centimeters. Perhaps this is in fact

CHART 1

Coevolution: Possible Cases

Physical Characteristic	Example	Putative Rationale(s) for Selection
Leg length	Africans versus Europeans versus Asians	1. Timing of puberty: Immediately prior to puberty, legs grow rapidly. Therefore populations in which puberty is delayed have extra time during which leg growth is paramount. 2. Heat loss per unit volume: Africans have long limbs so that heat loss per unit volume is high.
Lung size/chest circumference	Quechua children in high altitudes of Peru have larger lungs and chest circumference than do Quechua children living on the coast.	Relative richness of oxygen content of the atmosphere.
Sickle cell anemia (presence/absence of S allele, which causes a biochemical alteration in the structure of hemoglobin)	Much higher frequency of condition among those of African descent than those of non-African descent. Among West Africans, more frequent among yam cultivators than non-yam (rice, etc.) cultivators (malaria more common in yam-producing areas).	"Balancing" selection pressure of malarial mortality; resistance to malaria enhanced by presence of S allele.
Adult lactose absorption capacity (in other mammals lactose absorption capacity is limited to infants)	Adult lactose adsorption capacity most prevalent in populations with a long history of dairy production and/or a chronic deficiency related to incident ultraviolet light.	Cultural differences in frequency of dairying or the way milk is processed into food (e.g., yogurt versus drinking milk) may favor genetic evolution that allows for adult lactose absorption.

SOURCES: Durham 1991; Eveleth and Tanner 1990; Tanner 1978.

the case, and the failure to reach putative genetic potential is a fully satisfactory explanation. But at the present stage of our knowledge of auxology we simply do not know whether coevolution can be totally ruled out,
even for analysis covering a period as short as a century. Care must thus
be exercised in interpreting statistical associations between secular movements in components of the standard of living defined in opulence terms
and secular changes in height, weight, and related anthropometric measures. For simplicity in the analysis of secular trends in chapter 2, I will
not explicitly discuss coevolution. I will argue that changes in net nutritional intake are the dominant factor in accounting for changes in population quality in Japan. But I must again warn the reader that we cannot
completely simply dismiss the possibility that coevolution may be operating even over a period as short as eight and a half decades.

GROSS AND NET NUTRITION

By net nutritional intake, I mean total (gross) nutritional intake net of
the nutrients used to fuel physical and mental work and to fight off disease. In fashioning this definition I ignore the nutritional intake used up
in states of pure rest like sleeping (the so-called basal metabolic rate).[6]

The first of the two key hypotheses of this study is that a major cause
of the improvement in population quality in Japan over the 1900–1985
period is a secular improvement in net nutritional intake. To put the hypothesis in simple mathematical terms my claim is that

$$Q = f(N) \tag{1.1}$$

where f stands for some mathematical function, Q is an indicator of
population quality, and N is net nutrition. Now we can write net nutritional intake as gross nutritional intake GN minus the nutritional resources used up in staving off disease and in physical work. Let D stand
for an index of the incident of disease and L for the demands placed on
nutritional intake by physical labor. Then we can write

$$N = GN - \lambda D - \delta L \tag{1.2}$$

where λ and δ are parameters of negative value. Thus we can rewrite
equation (1.1) as

$$Q = g(GN, D, L) \tag{1.3}$$

where g is a mathematical function. I will devote chapter 2 to exploring
variants of equation (1.3) in terms of a variety of proxy variables for Q,
GN, D, and L.

ORGANIZATION OF THE STUDY

We can now, by way of setting the stage for the analysis presented in the remainder of this book, draw together the various themes touched on in this chapter. The view advanced in this study is that population quality has vastly improved in Japan because of improvements in net nutritional intake. When analyzing changes in national aggregate averages, my view is that equation (1.3) (or variants of it) suffices for a quantitative analysis of the relationships involved. This type of analysis is the burden of chapter 2. While the net nutrition hypothesis finds significant support in the analysis presented in chapter 2, it is incomplete because it brushes over the social and economic context within which demand for population quality interacts with supply factors like technological improvements in food production and in medicine level in determining actual outcomes for population quality. To explore demand, we must consider the way demand is voiced both through markets and through nonmarket mechanisms like the demand for entitlements that redistribute demand among various socioeconomic groups. That is, we must undertake cross-sectional analysis comparing regions and socioeconomic groups and we must consider the ways in which various groups and regions voiced their demands for entitlements. Bringing entitlements into the analysis forces us to consider community and government. Governments regulate and set standards for foodstuffs and medicines. Governments also provide public goods and affect the levels of entitlements enjoyed by individual households through various mechanisms of redistribution. But governments are not the only organizational entities that shape the entitlements available to individual households. Enterprises also set standards for their workers and provide entitlements for employees and their household members. In Part II I explore the development of the institutions affecting household entitlements over public health and medicine and foodstuffs and in doing so highlight the balkanization of entitlements in prewar Japan. The story I develop is very much the stuff of economic history since it stresses the strength of the market, and it is also the stuff of social history since it stresses social unrest aimed at voicing demand for entitlements. In short, recounting this story that turns on the social history of population quality in Japan underlines the point that we must never neglect the role of political and social factors in the shaping of the great secular trends in population quality evident for the industrialized nations over the last several centuries.

CHAPTER 2

Secular Trends in Anthropometric Measures of Human Growth and Their Relationship to Net Nutritional Intake

Here I take up the twin tasks of measuring and documenting the secular improvement in population quality in Japan over the period 1900–1985 and of examining the causal role played by improvements in net nutritional intake. The primary reason for focusing on this period is the availability of a long-term annual time series on height, weight, and chest girth for schoolchildren assembled by the Ministry of Education. Fortunately it also proved possible to construct time series for a variety of proxies for the factors underlying net nutritional intake over the same historical stretch of time: gross nutritional intake, the incidence of diseases likely to affect young persons, and the flow of child/youth labor services. The findings support the basic hypothesis of this study, that net nutritional intake is a major determinant of population quality.

I wish to stress, however, that I do not claim that the conclusions arrived at in this chapter have been proven in any absolute sense of that word. What I do show is that, using certain techniques and specific proxy variables, the hypothesis that net nutritional intake matters cannot (within reasonable bounds of confidence) be rejected. Thus the tests I use are based on what is usually known as the classical theory of statistical inference (see Maddala 1992; Pindyck and Rubinfeld 1991). By their very nature, the tests and evidence offered here are provisional: their validity and appropriateness depend on the choice of proxy variables, the degree to which the variables are subject to measurement error, the design of the statistical procedures themselves, and so forth. That I feel my efforts are not in vain flows from my view that empirical work in economic and social history is fundamentally different from that in fields where one can conduct an ongoing, although finite, number of repeated and controlled experiments. Study of a historical process ultimately relies on processing some portion of the extant record left from actual realized experience in the past. One cannot redo the experiment over and over again. Rather, one can find new data from other periods, other countries, or other sources for the particular country or group of countries currently being examined which have not yet been explored, or one can select different ways of processing the data already in the hands of the community of researchers. Therefore, the field proceeds through an inductive process in which rejection, acceptance, or partial acceptance of hypotheses is used to motivate data gathering and to condition future hypothesis testing.

The chapter is organized as follows: first, I document the secular changes in the anthropometric measures for schoolchildren which serve as the measures for various facets of population quality in Japan; second, I discuss series on nutrition, public health and medicine, and child/youth labor input that will serve as proxies for the factors underlying changes in net nutritional intake; third, I report the fruits of regression analysis using various econometric methods devised for the analysis of time series data. Because the regression techniques I use are probably not well known by all readers of this book and because some explanation must be given as to why I have selected the ones I have chosen, in an appendix to this chapter I discuss technical statistical issues in some detail.

THE SECULAR IMPROVEMENT IN HEIGHT, WEIGHT, AND CHEST GIRTH, 1901–1985

The anthropometric measures consist of a complex of related measures of physique: height, weight, weight for height (BMI, the body mass index, defined as weight in kilograms divided by the square of height calculated in meters), chest girth, and so forth. The most commonly utilized of the measures is (standing) height.[1] The widespread use of height as a measure of the standard of living in the field of anthropometric history reflects the fact that for the last several centuries figures on height have been collected and published by and for the use of military organizations in a number of countries (see panel A of table 2). The degree to which military data is representative of the terminal heights of young adult males is a matter of debate in particular cases (see the special methods developed by Floud, Wachter, and Gregory [1990] to handle British data of this sort). But the fact remains that even when one discounts for problems of measurement, there are clear secular trends in adult height for the United States and select European countries. Indeed the estimates marshaled in panel A of table 2 suggest that while heights for males in the United States did not rise appreciably, heights in the Scandinavian countries showed a dramatic increase. Figures for male schoolchildren aged eighteen in Japan show increases that in percentage terms are comparable to those for Scandinavia. For instance, comparing the figures in panel B.1 of table 2 for the years 1981–1985 with those for 1901–1985, it is apparent the percentage gain in height for males is about 6.6 percent, which is roughly equivalent to the percentage gains for Sweden and Norway recorded over a historical period that is almost twice as long.

To a degree the figures on height suggest that the population has "caught up" with Western Europe or at least closed the gap that once existed. However, one must be careful in asserting this. A gap still remains; on average in 1985 males in Japan tended to be shorter than males in most Western European countries. During the 1981–1985 period height for males in Japan averaged about 170 centimeters. These average levels were reached by males in select European countries and the United States at the following dates: United States, 1715; Sweden, 1913; Norway, 1927; Denmark, 1930; Holland, 1950; France, 1960; and Italy, 1977 (Floud, Wachter, and Gregory 1990:26). Whether this increase in heights in Japan will continue and whether the gap will ever be eliminated is, of course, a matter that we cannot resolve here. But one point that was raised in chapter 1 must be kept in mind: the gene pool has a definite effect on standing heights in Japan. Note from the figures in panel B.2 of table 2 that most (but not quite all) of the gain in standing height has been due to a gain in leg length. As can be seen, the ratio of sitting-to-standing height has been declining as legs have become longer.

The other point that can be readily gleaned from a perusal of the figures for Japan in table 2 is the dominance of secular change in tempo over secular change in level. I give figures on ages 6 and 12 as well as on age 18 both to capture the dynamics of the growth spurt and to take advantage of the six-year differences between the ages. Creating a six-year standard interval is of considerable utility to the statistical analysis

TABLE 2

Height for Young and Adult Males in Western Europe, the United States, and Japan, 1750-1985

A. Heights for Adult Males in Europe and the United States

A.1. Levels (cm)

Approx. Date	U.S.	U.K.	Sweden	Norway	Netherlands	France
1750	172	165	167	165	n.a.	n.a.
1800	173	167	166	166	n.a.	163
1850	171	166	168	166	164	167
1900	171	167	172	171	169	165
1950	175	175	177	178	178	170

A.2. Indexes, 1750 = 100

Approx. Date	U.S.	U.K.	Sweden	Norway
1800	100.6	101.2	99.4	100.6
1850	99.4	100.6	100.6	102.4
1900	99.4	101.2	103.0	103.6
1950	101.7	106.1	106.0	107.9

A.3. Indexes, 1850 = 100

Approx. Date	U.S.	U.K.	Sweden	Norway	France
1900	100.0	100.6	102.4	101.2	98.8
1950	102.3	105.4	105.4	105.3	101.8

TABLE 2 continued

B. Standing and Sitting Height for Males Ages 6, 12, and 18 and Gains in Height for Males Ages 6 to 12 and 12 to 18, 1901-1985

B.1. Standing Height, Sitting Height, and Gains in Standing Height

Period	Standing Height (cm)			Sitting Height (cm)			Gain, Standing Height (cm)		
	Age 6	Age 12	Age 18	Age 6	Age 12	Age 18	Ages 6-12	Ages 12-18	Ages 6-18
1901-1910	106.7	133.6	159.9	n.a.	n.a.	n.a.	27.2	26.8	54.3
1911-1920	106.9	134.4	160.8	n.a.	n.a.	n.a.	28.4	26.9	54.9
1921-1930	107.7	136.2	161.6	n.a.	n.a.	n.a.	29.8	26.2	55.5
1931-1940	108.8	138.2	162.9	62.3	75.1	88.9	30.8	25.1	54.1
1941-1950	108.5	138.4	162.9	62.1	74.1	88.3	28.6	25.2	57.2
1951-1960	110.3	139.3	165.0	62.8	75.7	89.9	32.4	27.4	57.7
1961-1970	113.4	144.9	167.7	64.0	78.3	90.3	34.1	23.6	55.9
1971-1980	115.3	148.6	169.0	64.7	79.5	90.0	34.4	21.5	55.7
1981-1985	116.1	149.9	170.4	65.1	79.9	90.3	n.e.	n.e.	n.e.

TABLE 2 *continued*

B.2. Gains in Sitting Height and Ratios of Sitting to Standing Height

Period	Gains, Sitting Height (cm)			Ratio, Sitting to Standing Height, Levels (%)			Ratio, Gains in Sitting to Standing Height (%)		
	Ages 6-12	Ages 12-18	Ages 6-18	Age 6	Age 12	Age 18	Ages 6-12	Ages 12-18	Ages 6-18
1901-1910	n.a.	n.a.	n.a.	n.a.	n.a.	n.a.	n.a.	n.a.	n.a.
1911-1920	n.a.	n.a.	n.a.	n.a.	n.a.	n.a.	n.a.	n.a.	n.a.
1921-1930	n.a.	n.a.	n.a.	n.a.	n.a.	n.a.	n.a.	n.a.	n.a.
1931-1940	n.a.	n.a.	n.a.	57.3	54.6	54.7	n.a.	n.a.	n.a.
1941-1950	13.8	15.6	28.2	57.0	54.6	54.3	45.0	53.0	48.8
1951-1960	14.5	14.5	27.5	56.9	54.4	54.5	45.9	53.0	47.7
1961-1970	15.1	11.9	26.0	56.5	54.0	53.9	44.3	50.3	46.4
1971-1980	15.1	10.8	25.9	56.1	53.5	53.2	43.9	50.0	46.5
1981-1985	n.e.	n.e.	n.e.	56.1	53.3	53.0	n.e.	n.e.	n.e.

SOURCES: Steckel 1994b: table 7; Japan Statistical Association 1988: tables 21-3-a (pp. 122-125) and 21-3-d (pp. 134-135).

NOTES: Figures for sitting height given for 1931-1940 are actually in the case of ages 12 and 18 for 1937-1938 and in the case of age 6 for 1937-1939. Figures for sitting height given for 1946-1950 are actually for 1949-1950. Gains in height are calculated for year t by subtracting the value of height for children aged x in year t from the value of height for children aged $x + 6$ in year $t + 6$. Figures for 1921 and 1975 estimated by averaging values for surrounding years.

n.a. = not available.

n.e. = not estimated.

taken up in section 2.4, but an additional benefit can be immediately grasped here. We can calculate the gain in average height for persons aged 12 in a given year t and the same cohort of persons aged 6 in year t – 6. That is, we take the average height for persons aged 12 in year t and subtract from that the average height for persons aged 6 in year t – 6 in order to estimate the gain for the cohort in year t – 6. A secular trend toward earlier maturation, that is, a downward drift in the mean age of the growth process and in the adolescent growth spurt in particular, can be seen from the fact that the anthropometric measures at age 6 and age 12 increase disproportionately in comparison to the gains at age 18. For instance, the percentage gains in height over the 1901–1910/1981–1985 period at age 6 is 8.8 percent and at age 12 is 12.2 percent. The fact that mean age of the growth spurt is shifting downward means that the secular trend in the gains in the anthropometric measures between age 6 and 12 is unusually great. For instance, the age 6 to age 12 gain in centimeters averaged over the 1901–1910 period is 27.2 and the gain averaged over the 1971–1980 period is 34.4. In short, the secular trend in the tempo of growth in height up to age 6 dominates over the secular trend in levels of height at age 18. That is to say, the improvement in population quality can be discerned both in the trends in levels and in the trends in tempo.

The dominance of secular trends in tempo over secular trends in levels is even more evident for females than it is for males. Figures are given in table 3. The percentage increases in height levels over the 1901–1910/1981–1985 period for girls aged 6, 12, and 18 are 9.3 percent, 12.6 percent, and 6.4 percent, respectively. The percentage gain at age 18 for females is less than that for males at age 18, but the percentage gains at age 6 and at age 12 exceed the gains for males. The difference in secular changes in tempo is surely linked to the fact that females mature at younger ages than do males, which can be easily gleaned by comparing the gains between ages 6 and 12 for females with those for males. Note that the increase in the amount of the gains from ages 6 and 12 between 1901–1910 and 1971–1980 is identical for both sexes (7.2 cm) and that the gains are always larger for females than for males. Note also that gains between ages 12 and 18 are far larger for males than for females. This reflects both earlier maturation for females and the fact that terminal average adult heights for females are less than those for males. The secular decrease in gains in height for females aged 6 and 12 is an unusually dramatic indicator of the secular trend in the tempo of growth.

TABLE 3

Standing and Sitting Height for Females Aged 6, 12, and 18 and Gains in Height, Ages 6 to 12 and 12 to 18, 1901-1985

A: Standing Height, Sitting Height, and Gains in Standing Height

Period	Standing Height (cm)			Sitting Height (cm)			Gains, Standing Height (cm)		
	Age 6	Age 12	Age 18	Age 6	Age 12	Age 18	Ages 6-12	Ages 12-18	Ages 6-18
1901-1910	105.6	133.8	148.0	n.a.	n.a.	n.a.	29.0	15.1	44.1
1911-1920	105.5	135.2	149.4	n.a.	n.a.	n.a.	30.8	14.9	45.2
1921-1930	106.3	137.4	150.4	n.a.	n.a.	n.a.	32.5	13.9	46.0
1931-1940	107.8	139.7	152.0	61.5	76.1	84.3	33.3	13.1	45.2
1941-1950	107.6	139.7	152.9	61.7	75.1	84.1	31.1	13.7	46.8
1951-1960	109.5	141.1	154.1	62.4	77.5	84.3	35.1	13.9	46.6
1961-1970	112.4	146.5	155.8	63.5	80.2	84.9	36.3	9.9	44.3
1971-1980	114.4	149.7	156.6	64.1	81.2	84.2	36.2	7.6	43.6
1981-1985	115.4	150.7	157.5	64.6	81.5	84.1	n.e.	n.e.	n.e.

TABLE 3 continued

B. Gains in Sitting Height and Ratios of Sitting to Standing Height

Period	Gains, Sitting Height (cm)			Ratio, Sitting to Standing Height, Levels (%)			Ratio, Gains in Sitting to Standing Height (%)		
	Ages 6-12	Ages 12-18	Ages 6-18	Age 6	Age 12	Age 18	Ages 6-12	Ages 12-18	Ages 6-18
1901-1910	n.a.	n.a.	n.a.	n.a.	n.a.	n.a.	n.a.	n.a.	n.a.
1911-1920	n.a.	n.a.	n.a.	n.a.	n.a.	n.a.	n.a.	n.a.	n.a.
1921-1930	n.a.	n.a.	n.a.	n.a.	n.a.	n.a.	n.a.	n.a.	n.a.
1931-1940	n.a.	n.a.	n.a.	57.0	54.5	55.4	n.a.	n.a.	n.a.
1941-1950	16.0	9.3	23.1	57.2	54.8	55.1	47.5	53.8	49.2
1951-1960	17.0	7.2	22.6	57.0	54.9	54.7	48.4	52.2	48.5
1961-1970	17.4	4.4	20.6	56.5	54.8	54.5	48.0	44.2	46.4
1971-1980	17.3	2.9	20.1	56.1	54.2	53.8	47.8	38.5	46.1
1981-1985	n.e.	n.e.	n.e.	56.0	54.1	53.4	n.e.	n.e.	n.e.

SOURCES: Japan Statistical Association 1988: tables 21-3-a (pp. 122-125) and 21-3-d (pp. 134-135).

NOTES: Figures for sitting height for 1931-1940 actually are for 1937-1938 in the case of ages 12 and 18 and for 1937-1939 in the case of age 6. Figures for sitting height for 1946-1950 actually are for 1949-1950. Gains in height were calculated for year t by subtracting the value of height for children aged x in year t from the value of height for children aged x + 6 in year t + 6. Figures for 1921 and 1975 were estimated by averaging values for surrounding years.

n.a. = not available.

n.e. = not estimated.

While the differences between males and females in terms of the tempo of growth in and levels of height are of interest, the main point I wish to stress is the similarity in the secular trends between the sexes. The correlations between levels of male and levels of female height at ages 6, 12, and 18 are very high: over the 1900–1985 period the correlations are +.997, +.99, and +.97, respectively; and over the 1900–1940 period the correlations are +.97, +.98, and +.96, respectively.

That trends in tempo and level differ somewhat in magnitude does not mean that the heights at the various key ages that we focus on here are not highly correlated. For instance, over the 1900–1985 period the correlation between heights for males at age 6 and 12 is +.98; at ages 6 and 18, +.99; and at ages 12 and 18, +.97. For females the correlations over the same period are somewhat lower: +.96, +.90, and +.91, respectively. That correlations are lower for females probably reflects the striking dominance of secular trend of tempo. It is interesting and useful to keep in mind that over the period 1900–1940 the correlations between heights at ages 6 and 12, at ages 6 and 18, and at ages 12 and 18 are lower than over the entire period 1900–1985. For instance, for males the correlations are respectively +.96, +.90, and +.91; and for females the correlations are respectively +.91, +.88, and +.92. A final point to keep in mind in considering both the differences and similarities between secular trends in the tempo and in the levels of human growth is the greater sensitivity of six-year gains in height to changes in nutrition and other components of per capita consumption. To see this consider figure 1, which graphs annual figures on an index for nutrition (described in the next section), heights for females aged 6, and six-year gains in height between ages 6 and 12. Notice that the drop in nutritional intake occurring in the late 1930s and early 1940s due to Japan's growing military involvement is clearly mirrored with a slight lag in the case of six-year gains in height but that the impact on height levels is far less pronounced. In sum, tempo dominates over levels in terms of secular trend; it is also a more sensitive barometer of changes in the factors underlying population quality. To be sure, individuals who are deprived in the short run may "catch up" later on and therefore the overall impact of a short-term diminution in net nutrition may not be long lasting. Nevertheless, insofar as we are interested in ascertaining the causal connection between net nutrition and population quality, the greater sensitivity of gains in height and gains in other anthropometric measures is an important point that must not be forgotten.

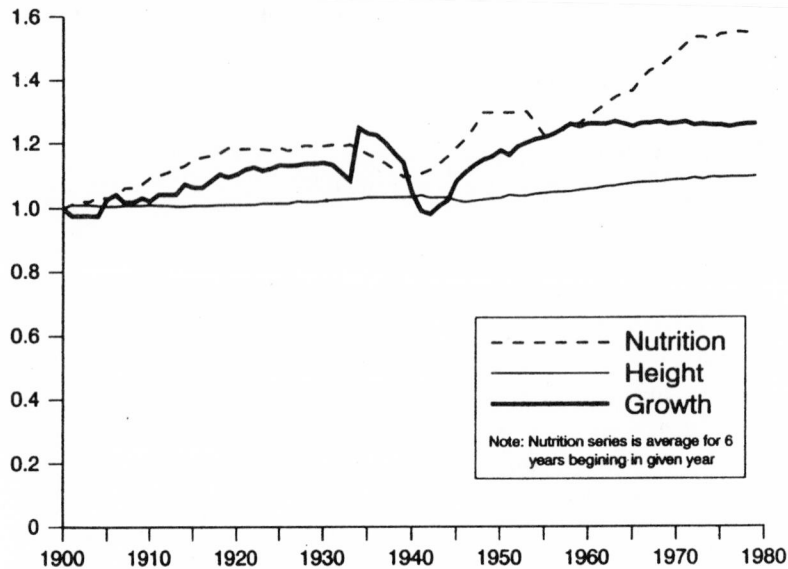

Figure 1.   Indexes of Nutrition, Height for Females Age 6 and Growth for
Females from Age 6 to 12

I have begun my discussion of secular trends in the anthropometric
measures in Japan with figures on height because height figures are the
ones that are most commonly encountered in anthropometric history.
While I do make use of some Japanese military conscription data later
on in this study and T. Shay (1994) makes extensive use of it in his
study, the bulk of my data is for schoolchildren and comes from surveys
conducted in schools by the Ministry of Education. Why do I favor the
use of this data? First, it is more comprehensive than the military recruitment examination data: it covers both sexes; it includes figures on
weight and chest girth; and it is available throughout both the 1900–
1940 and postwar periods (the military draft in Japan ended after
World War II, although Japan currently maintains a small self-defense
force). Second, the data are available for a variety of ages under 18 and
hence allow analysis of tempo effects that, as noted above, are far more
sensitive to short-run fluctuations in net nutrition than are levels, especially adult levels.

Despite the compelling virtues of the data set analyzed here, there are
defects in it which we must confront before proceeding further. Chart 2
provides a summary background for a general discussion of the data
used not only in this section but throughout this chapter. At this junc-

ture I will concentrate on the anthropometric measures covered in panels A, B, and C of the chart. As can be seen, the figures are for schoolchildren, and therefore children in the relevant age groups who did not attend school are excluded. Who was attending school? This depends on the period involved. S. B. Levine and H. Kawada (1980: 48–52) note that by 1900 four years of schooling (typically from ages 6 to 10), made compulsory in 1886, was virtually universal in practice. In 1907 it was decreed that compulsory education be extended to six full years, although this requirement was not successfully enforced until 1918. However, the extent of underregistration should not be exaggerated. For instance, in 1910, 98.8 percent of males and 97.4 percent of females of compulsory school age were reported as attending school (Japan Statistical Association 1988: table 22–1, p. 212). It is thus reasonable to suppose that by the early 1900s most children aged 6 and 12 are covered in the Ministry of Education data set. After the war nine years of education was made compulsory (from ages 6 to 15), and by the 1970s most children in Japan were in fact graduating from high school, which means that the coverage of individuals aged 18 in the Ministry of Education data set is by and large universal for all three ages analyzed here. The problem, of course, is the potential bias resulting from the selectivity of enrollment of children over age 12 before that period.

As can be seen from panel A of chart 2, there is a potential upward bias in heights, weights, and chest girth for males and females aged 18; and the earlier the date, the greater the upward bias. The reason is that the earlier the date, the lower the advancement rate past age 12 in the school system, and as a result the greater the selectivity of the population of those examined by the Ministry of Education. It is demonstrated in Part II of this volume that there are socioeconomic differentials in the anthropometric measures: in general the higher the social and economic status one is born into, the greater the height, weight, and chest girth and the higher the probability of advancing past age 12 in the school system. But as advancement rates increased the extent of this upward bias began to fall. It is likely that there is a contrary tendency at work on heights, weights, and chest girths for individuals aged 18, a contrary tendency that may allow us to conclude that this declining upward bias does not present a major problem for our analysis. The reason for the existence of a contrary tendency lies in the declining mean age of maturation. While I use age 18 as my oldest age for analysis (since advancement rates in the educational system after age 18 are so low that we

CHART 2

Notes on Estimates of Anthropometric Measures and Indexes of Nutrition, Public Health and Medicine, and Child/Youth Labor Input Used in Time Series Analysis and Selected Correlations for Anthropometric Measures of 18-Year-Olds

A. Anthropometric Measures, Notes

Variable	Nature of Series	Comments	Abbreviations
Standing height and gain in standing height	For males and females ages 6, 12, and 18. Gains in height calculated by taking the difference between heights for individuals 6 years older 6 years later and the present heights in the present year.	Underlying data available for the years 1900-1985. Data missing in 1921 and 1947 and estimated for those years by taking averages for surrounding years. Data collected by the Ministry of Education with the School Examination Survey and the Physical Fitness Test. Beginning in 1968, working youths surveyed. There is an upward bias in the heights for 18-year-olds which can be surmised by comparing figures in tables 2, 4 and 7.	H, GH
Weight and gain in weight	See discussion above for height.	See discussion above for height.	W, GW
BMI	See discussion above for height.	See discussion above for height.	BMI, GBMI
Chest girth (CG)	See discussion above for height.	See discussion above for height.	CG, GCG

B. Correlations Between Various Anthropometric Measures for 18-Year-Olds: Heights and Weights

Measure	1900-1985 (HM = Male height, etc.)				1900-1940				1948-1985			
	HM	HF	WM	WF	HM	HF	WM	WF	HM	HF	WM	WF
HM	+1.00	+.97	+.98	+.85	+1.00	+.96	+.97	+.93	+1.00	+.97	+.98	+.74
HF		+1.00	+.95	+.91		+1.00	+.97	+.93		+1.00	+.97	+.76
WM			+1.00	+.85			+1.00	+.94			+1.00	+.72
WF				+1.00				+1.00				+1.00

CHART 2 *continued*

C. Correlations Between the BMI for 18-Year-Olds

Measure	1900-1985 (BMIM = Male BMI, etc.)		1900-1940		1948-1985	
	BMIM	BMIF	BMIM	BMIF	BMIM	BMIF
BMIM	+1.00	-.41	+1.00	-.32	+1.00	-.66
BMIF		+1.00		+1.00		+1.00

D. Index for Public Health and Medicine: Notes

Variable	Nature of Series	Comments	Abbreviations
Index for public health and medicine	[1] Let DOCPC = doctors per 100,000 population and let DOCPCI = index for DOCPC with 1900-1904 = 100. [2] Let DI = combined death from 4 major infectious causes (tuberculosis, pneumonia, bronchitis, and enteritis) and let DIR be the death rate (per 100,000 population) from these causes. Then define IDIR = 1/DIR as the inverse death rate and let IDIRI be the index with 1900-1904 = 100 for IDIR. [3] Let CPDR = cases per death for 4 causes (cholera, dysentery, typhoid fever, and smallpox) and let CPDRI be the index for CPDR with 1900-1904 = 100. [4] Then PHMEDI = 1/3(DOCPCI) + 1/3(IDIRI) +1/3(CPDRI).	Available for 1900-1985. Standards for certification of doctors changed in the early twentieth century when a knowledge of Western medical techniques became a prerequisite for certification.	PHMEDI, PHMEI, PHMI, PH

E. Indexes for Child/Youth Labor Input

Variable	Nature of Series	Comments	Abbreviations
Index of child/youth labor input #1	For each sex separately: [1] Let PPI = % gainfully employed who are in primary industry and PPII be the index based on PPI with 1900-1904 = 100; and [2] Let LFPR = labor force participation rate for individuals aged 10-19 and LFPRI be the index based on LFPR with 1900-1904 = 100; then [3] The ICYI1 = index for child/youth labor input = 1/2(PPII) + 1/2(LFPRI).	Data are available for 1900-1985. Precise calculation of rates is rendered difficult by the existence of multiple job holding, that is, the holding of secondary jobs along with primary jobs (e.g., farmers working in factories during the week and on their farms during the weekends). The series for males and females are separate.	ICYI1, ICYII1
Index of child/youth labor input #2	In addition to the two variables PPII and LFPRI considered above, the second index incorporates the index (with 1948-1950 = 100) for the % of workers who are not employees (PWNEI). ICYI2 = 1/3(PPII) + 1/3(LFPRI) + 1/3(PWNEI).	See comments above for ICYI1. Series is available for 1948-1985.	ICYI2, ICYII2

CHART 2 continued

F. Indexes for Nutrition

Variable	Nature of Series	Comments	Abbreviations
Index of nutrition #1	All nutritional series are per day/per capita and all indexes are with 1900-04 = 100: [1] Let CAL be calories (in kcal) and be CALI be the index for CAL; [2] Let PRO be protein (in grams) and let PROI be the index for PRO; [3] Let VITA be vitamin series in international units and VITAI the index; and let VITB1 be the vitamin B1 series in milligrams and VITB1I the index; let VITB2 be the vitamin B2 series in milligrams and let VITB2I be the index; and let VITC be the vitamin C series in milligrams and VITCI the index; then weighting each of the four vitamin indexes by 1/4 and adding calculate VITI. [4] The overall index NUT1I = (.4)CALI + (.4) PROI + (.2) VITI.	Data are available for the entire period 1900-1985. The postwar estimates were constructed by the Ministry of Health and Welfare. From fiscal 1946 through fiscal 1964 these estimates were unweighted averages of survey results obtained four times a year. None of the data is adjusted for nutrient loss in cooking. The method of estimation for calorie intake changed in the late 1960's and the method for calculating vitamin A changed in 1955.	NUT1, NUT1I
Index of nutrition #2	[1] Let CALC be calcium intake (in milligrams) and let CALCI be the index based on the series with 1946-50 = 100; and [2] Let FAT be fat intake (in grams) and let FATI be the series with 1946-1950 = 100. [3] Then CALCFAI = (.5) CALCI + (.5) FATI.	Data are available from 1946 to 1985. There is a fairly high correlation between an index for protein intake and CALCFAI throughout this period.	CALCFAI
Index of nutrition #3	In addition to the nutrition series included in the two indexes above, the following series (converted to indexes with 1946-1950 = 100) are included in the most comprehensive index: carbohydrates (in grams with index CARBI) and iron (in mg with index IRONI). Then the overall index NUT2I = (.2) CALI + (.2) PRO + (.1) FATI + (.1) CARBI + (.1) CALCI + (.1) IRONI + (.2) VITI.	Data are available for 1946 to 1985 with the exception of the carbohydrate series which began in 1949. I assumed carbohydrate intake for 1946-1948 was equal to the carbohydrate intake in 1949. There are no data on iron intake between 1964 and 1970 and I assumed it was equal to that for the average of iron intake in 1963 and 1971.	NUT2, NUT2I

cannot usefully employ data on persons over 18 as auxological indicators for the population at large), in fact growth does not necessarily terminate at age 18. But as the mean age of maturation declines, the proportion of the population that continues to mature after age 18 declines. Hence it is likely that the secular trend toward earlier maturation, according to which there is a declining downward bias in the anthropometric measures (the earlier the year, the greater the downward bias), does counteract the upward bias due to a decline in the selectivity of the population receiving an education through age 18. For this reason I feel relatively comfortable using data for all three ages, 6, 12, and 18. But I must caution the reader that the problem does exist.

Now to return to a point made earlier about the virtues of using a data set that allows us to measure and analyze a variety of anthropometric characteristics of Japan's population, let us turn to figures on weight and the BMI. Figures for males appear in table 4 and figures for females appear in table 5. The virtues of having data on weights for both males and females (not just males, as is usually the case when one is using military recruitment data) can be seen from a comparison of trends in the BMI for males and for females. For if we calculate percentage gains in weight for the period 1901–1910 to 1981–1985, we get the following figures for percentage increase in weight and the BMI.

	Weight	BMI
Males		
Age 6	+20%	+1.3%
Age 12	+40%	+10.8%
Age 18	+18%	+3.9%
Females		
Age 6	+22.6%	+2.7%
Age 12	+40%	+10.6%
Age 18	+7.6%	−5.3%

Note that the increase in weight for females at age 18, contrary to the increase at ages 6 and 12, is far less than that for males. The reason cannot be physiological: as can be seen from table 5, this is almost entirely a postwar phenomenon. The obvious explanation is dieting, and the reason for dieting is the concept of beauty for women that places strong emphasis on being slender. (The concept of sacrificing potential physical work capacity for beauty is undoubtedly much more typical of the urban nonagricultural population than of the farming population for whom potential for work is a virtue among young women on the

TABLE 4

Weight and the Body Mass Index: Levels and Gains in Levels, Males, Ages 6, 12, and 18, 1901-1985

A. Weight and Gains in Weight

Period	Weight (kg)			Gain in Weight (kg)		
	Age 6	Age 12	Age 18	Ages 6-12	Ages 12-18	Ages 6-18
1901-1910	17.5	29.8	52.3	12.5	23.0	35.7
1911-1920	17.6	30.2	53.1	13.3	23.3	36.5
1921-1930	17.7	31.4	53.8	14.4	23.2	37.6
1931-1940	18.2	32.7	55.0	15.0	22.9	36.6
1941-1950	18.3	32.6	55.0	13.9	22.1	37.9
1951-1960	18.7	33.3	55.7	16.5	23.7	39.7
1961-1970	19.6	36.7	57.9	19.4	22.5	40.6
1971-1980	20.5	40.2	60.0	20.7	21.1	41.7
1981-1985	21.0	41.6	61.7	n.e.	n.e.	n.e.

B. BMI and Gains in BMI

Period	BMI			Gain in BMI		
	Age 6	Age 12	Age 18	Ages 6-12	Ages 12-18	Ages 6-18
1901-1910	15.4	16.7	20.5	1.4	3.8	5.2
1911-1920	15.4	16.7	20.6	1.5	3.9	5.3
1921-1930	15.3	16.9	20.6	1.7	3.8	5.5
1931-1940	15.4	17.1	20.7	1.7	3.7	5.3
1941-1950	15.5	17.0	20.7	1.6	3.5	5.0
1951-1960	15.4	17.2	20.7	1.9	3.4	5.3
1961-1970	15.3	17.5	20.6	2.7	3.4	5.7
1971-1980	15.5	18.2	21.0	3.0	3.0	6.0
1981-1985	15.6	18.5	21.3	n.e.	n.e.	n.e.

SOURCES: Japan Statistical Association 1988: tables 21-3-a and 21-3-b (pp. 122-129).

NOTES: Figures for 1921 and 1975 estimated by averaging the figures for the surrounding years.

n.e. = not estimated

TABLE 5

Weight and Body Mass Index: Levels and Gains in Levels, Females, Ages 6, 12, and 18, 1901-1985

A. Weight and Gains in Weight

Period	Weight (kg)			Gain in Weight (kg)		
	Age 6	Age 12	Age 18	Ages 6-12	Ages 12-18	Ages 6-18
1901-1910	16.8	30.5	47.6	13.9	17.6	31.8
1911-1920	16.9	31.0	48.4	14.7	17.7	32.1
1921-1930	17.2	32.3	48.9	16.3	17.1	33.0
1931-1940	17.6	34.0	49.8	16.9	16.6	33.0
1941-1950	17.7	33.7	50.7	15.9	16.2	32.2
1951-1960	18.2	35.2	49.7	19.3	15.3	32.8
1961-1970	19.1	38.9	50.9	21.8	12.2	31.9
1971-1980	20.0	41.8	51.0	22.4	9.4	31.4
1981-1985	20.6	42.7	51.2	n.e.	n.e.	n.e.

B. BMI and Gains in BMI

Period	BMI			Gain in BMI		
	Age 6	Age 12	Age 18	Ages 6-12	Ages 12-18	Ages 6-18
1901-1910	15.1	17.0	21.7	1.9	4.7	6.6
1911-1920	15.2	17.0	21.7	1.8	4.6	6.4
1921-1930	15.2	17.1	21.6	2.1	4.5	6.4
1931-1940	15.1	17.4	21.6	2.2	4.3	6.4
1941-1950	15.3	17.3	21.7	2.2	3.9	5.7
1951-1960	15.2	17.7	20.9	2.7	3.4	5.7
1961-1970	15.1	18.1	21.0	3.4	2.8	5.6
1971-1980	15.3	18.7	20.8	3.4	2.0	5.3
1981-1985	15.5	18.8	20.6	n.e.	n.e.	n.e.

SOURCES: Japan Statistical Association 1988: tables 21-3-a and 21-3-b (pp. 122-129).

NOTES: Figures for 1921 and 1975 estimated by averaging the figures for the surrounding years.
n.e. = not estimated.

marriage market. Hence it is not surprising that the trend is much more evident during the postwar period when the farming population has been rapidly dwindling than during the prewar era.) Thus the weight and BMIs of young adults were apparently shaped by a cultural rule that does not appear to have been operating for younger children. This is an additional argument in favor of using data for young children as well as adults. It should be noted, however, that height of females at age 18 appears to be unaffected by the practice of dieting. Correlations between measures of height and weight for males and females presented in panels B and C of chart 1 testify to differences between height, on the one hand, and weight and the body mass index, on the other, which must be kept in mind in interpreting the anthropometric measures as measures of population quality.

Trends for a third major auxological measure, chest girth, are explored in table 6. Particularly striking for females is the secular trend in tempo that overshadows the secular trend in levels. For instance, in comparison with 1901–1910 when the six-year gain in chest girth for girls aged 6 is 11.6 centimeters, the average six-year gain during the 1981–1985 period is 18.2 centimeters.

Finally, the figures on anthropometric measures in table 7 illustrate several points about the drawbacks and strengths of the data set for school children on which my analysis in this chapter rests and provide additional background about prewar trends in height.[2] The following observation supportive of use of the data on schoolchildren leaps out at us. Trends in average height and weight for persons examined for conscription examinations are much less striking—indeed, the trends in averages are almost ambiguous—than are the trends we secured for schoolchildren. To be sure, it may be argued that truth may lie on the side of an ambiguous or uncertain trend and that, therefore, finding this trend in the military recruit data is a virtue and not a deficiency. But there are reasons for thinking that the figures on schoolchildren more accurately represent the true underlying pattern. To see my point, note that the figures on percentages "small" (defined as 150 cm or less) and on percentages "tall" (defined as 170 cm or more) point toward an unfaltering increase in height that is consistent with what—using averages—we have found for male schoolchildren. Indeed, the fact that moments of the distribution of heights other than averages are available for military recruits is one of the most attractive features of the military recruitment data. But the shortness of the time period for which the data on distribution of heights is available (1915–1940), the lack of figures

TABLE 6

Chest Girth and Gain in Chest Girth, 1901-1985

A. Males

Period	Chest Girth (cm)			Gain in Chest Girth (cm)		
	Age 6	Age 12	Age 18	Ages 6-12	Ages 12-18	Ages 6-18
1901-1910	54.0	65.4	80.5	11.7	15.5	27.4
1911-1920	54.1	65.9	81.2	12.0	15.9	28.2
1921-1930	54.4	66.2	82.1	12.1	16.5	29.3
1931-1940	54.7	66.9	83.3	12.8	17.1	28.4
1941-1950	55.7	67.4	83.5	11.8	15.4	28.3
1951-1960	56.3	68.1	83.5	12.7	16.8	29.7
1961-1970	56.8	69.9	85.7	14.5	16.3	29.5
1971-1980	56.8	71.9	86.3	15.6	14.8	29.6
1981-1985	57.7	72.6	86.7	n.e.	n.e.	n.e.

B. Females

Period	Chest Girth (cm)			Gain in Chest Girth (cm)		
	Age 6	Age 12	Age 18	Ages 6-12	Ages 12-18	Ages 6-18
1901-1910	52.5	63.7	77.5	11.6	14.6	26.4
1911-1920	52.3	64.6	78.7	12.8	14.2	25.9
1921-1930	52.6	65.4	78.4	13.5	12.9	26.6
1931-1940	53.1	66.8	78.8	14.8	13.1	27.3
1941-1950	54.2	67.8	80.3	13.5	12.8	26.2
1951-1960	54.7	68.9	80.2	15.8	12.1	26.7
1961-1970	55.3	71.7	81.3	18.1	9.7	26.1
1971-1980	56.1	74.0	81.4	18.2	7.5	25.0
1981-1985	56.3	74.4	81.5	n.e.	n.e.	n.e.

SOURCES: Japan Statistical Association 1988: tables 21-3-c (pp. 130-133).

NOTES: Figures for 1921 and 1975 estimated by averaging the figures for the surrounding years.

TABLE 7

Anthropometric Measures for 20-Year-Old Males Based on Military Conscription Examination Data, 1900-1940

Period or Year	% Small (150 cm or less)	% Tall (170 cm or more)	Average Height (cm)[a]	Average Weight (kg)[a]	BMI[a]
1900	16.7	1.3	n.a.	n.a.	n.a.
1901-1910	14.2	1.7	n.a.	n.a.	n.a.
1911-1920	11.4	2.4	160.1	51.9	20.3
1921-1930	7.2	3.6	159.5	52.3	20.6
1931-1940	3.8	4.7	160.3	53.0	20.6

SOURCES: Japan Statistical Association 1988: reference tables 21-1 and 21-2 (pp. 196-197).

NOTES: [a]Averages for 1915-1920.
n.a. = not available.

on females, the absence of any measurements capturing the timing of the growth spurt, and so forth, make sole reliance on these data alone problematic. I do make use of them to a limited degree in Part II of this study.

INDEXES OF NUTRITION, PUBLIC HEALTH AND MEDICINE, AND CHILD/YOUTH LABOR INPUT

The goal underlying construction of the measures discussed here is to secure proxy variable(s) capturing the three main factors underlying levels of net nutritional intake: gross nutritional intake and the uses to which human bodies put that nutritional intake other than in physical growth—physical exertion and work and fighting off diseases, especially infectious disease. The human body must either dip into its reserves of fat and muscle tissue or into flows of new nutritional intake, thereby potentially diverting the body's built-up, or freshly secured, reserves away from physical and mental growth, in generating energy that keeps it healthy and capable of sustaining physical exertion. The basic equation (1.3) decomposing gross nutritional intake into components that are used in human growth, into components that are utilized in fighting off disease, and into components that are consumed in physical work guides us in constructing the proxy variables. The extent to which the proxy variables capture or fail to capture the true factors underlying net nutritional intake for the average young person living in Japan during the 1900–1985 period is a matter that can be debated, but I believe my measures can be usefully employed in the time series analysis described in the next section.

The main assumptions underlying the proxy variables for the three components of gross nutritional intake are reviewed in chart 1. Here I will restrict my remarks concerning the variables to the following points concerning the philosophical approach that guided me in constructing the variables. One of my guiding principles was to construct variables that do not decisively depend on any one type of data or on any one type of assumption. For instance, it has been pointed out to me that there may be a bias in the figures on labor force participation for children and young adults used in the construction of the index of child/youth labor input during the first decade or so of the twentieth century (presumably the bias diminishes with time).[3] Since the labor force participation rate only receives half of the weight in the overall index, this potential source of bias is reduced, albeit not completely eliminated. Another guiding principle was, wherever possible, to construct two or more proxy variables for the underlying factor to see to what extent different assumptions about the appropriateness of a given proxy are valid. Hence in the case of gross nutritional intake I offer one series for the prewar period 1900–1940 and three for the postwar period up to 1985. Finally, since I am interested in relative and not absolute levels of the proxy variables, I constructed indexes (typically with the average value for 1900–1904 = 100) for each of the components of each overall index and then, by weighting each component index, calculated the sum of the components to arrive at an overall index. By construction, the resulting variable is a composite index.

Of the many potential problems that reduce the reliability of the series advanced here, one seems to me to be of sufficient concern to mention here: most of the series are per capita averages and hence may not accurately reflect the circumstances specific to children and young adults. This objection is not valid in the case of the proxy variable for child/youth labor input but is valid for the nutritional intake and public health and medicine series. Without doubt we do not know how food and public health and medical services were distributed within families. For instance, a bias in favor of males or eldest male sons among siblings has been extensively discussed in the literature on the Japanese family. Such a bias may exist, but the high correlations between the anthropometric measures for male and female children should be food for thought for those who would argue that such a bias undermines the validity of the measures used here. Moreover, it should be pointed out that experiments with dividing the nutrition series by a consumer unit weighted population—that is, by age- and sex-specific population weighted by numbers giving the relative consumption demand of the group compared to prime age males—did not reveal a significant difference in trends in nutrition between the per capita and the per consumer weighted population estimates.[4]

A useful summary of the various component indexes underlying the gross nutritional intake variable appears in table 8. Since some data were available only for the postwar period (e.g., estimates of carbohydrate and fat and calcium intake), two nutritional indexes, a composite index with wide coverage (NUTI2) and an index for calcium and fat intake (CALCFATI), are calculated for the postwar years only. The index available for the entire 1900–1985 period, NUTI1, is a composite of indexes for per capita per diem consumption of calories, protein, and four types of vitamins (see panel F of chart 2 and notes to table 8). Underlying

TABLE 8

Indexes of Daily Per Capita Nutritional Intake, 1901-1985

A. Indexes Covering Entire Period

Indexes for Calories (CAL1); Proteins (PROI); Vitamins A, B1, B2, and C Combined (VITI);
and Overall Nutrition (NUTI1), 1900-1904 = 100

Period	CAL1[a]	PROI	VITI[b]	NUTI1[c]
1901-1910	103.1	99.8	104.7	102.1
1911-1920	111.0	110.3	116.5	111.9
1921-1930	114.5	119.3	123.2	118.2
1931-1940	111.1	120.6	130.1	118.7
1941-1950	98.6	103.2	165.9	113.9
1951-1960	100.9	112.2	211.6	127.5
1961-1970	104.7	119.7	214.6	132.7
1971-1980	106.2	130.0	285.9	151.7
1981-1985	102.1	128.9	313.0	155.0

TABLE 8 *continued*

B. Separate Postwar Nutritional Indexes

Indexes for Calcium (CALCI), Fat (FATI), Calcium and Fat Combined (CALCFATI), and Overall Nutrition (NUTI2)

Period	CALCI	FATI	CALCFATI[d]	NUTI2[e]
1946-1950	74.8	78.3	76.5	95.9
1951-1960	110.6	110.9	110.7	97.4
1961-1970	145.1	192.0	168.6	108.5
1971-1980	166.1	278.4	222.3	128.7
1981-1985	170.6	297.8	234.2	131.6

SOURCES: Mosk and Pak 1978: various tables; Japan Statistical Association 1988: table 21-1 (p. 117).

NOTES: [a]The method of measuring calorie intake changed during the later 1960s, making comparison of pre- and post-1965 calorie data somewhat difficult. For 1941-1945 I estimated the calorie intake by assuming a constant annual amount of decline between 1940 and 1946.

   [b]The measurement of vitamin A changed in 1955. I assumed the values of vitamin A intake for the years 1946-1954 were equal to the 1955 value. And I assumed the values for 1941-1945 were equal to the 1940 values for all four types of vitamins. To construct the overall index for vitamin intake, I weighted each separate vitamin index by .25 and added the weighted indexes.

   [c]To construct the overall index, I weighted each of the calorie and protein indexes by .4 and added that total to the vitamin index weighted by .2 (calorie and protein intake each get a weight of .4).

   [d]To calculate the combined index, I weighted each of the calcium and fat indexes by .5 and added them.

   [e]To calculate the combined index, I used the following weights: calories, .2; protein, .2; fat, .1; carbohydrates, .1; calcium, .1; iron, .1; and vitamins combined, .2. None of the indexes take into account losses of nutrients in cooking.

the trends in all three indexes are two factors: increases in consumption of foods traditionally eaten during the preindustrial era before the 1880s, that is, an expansion in the quantity of nutrient intake with a given diet; and a change in the composition of food intake, that is, a structural shift in diet. As far as the composition of diet is concerned, it is important to recognize that the opening up of Japan to trade in the 1850s and the American Occupation after 1945 both left a mark on the dietary intake of the population of Japan. In particular, a shift toward calcium and fat partly because of a growing consumption of dairy products is directly related to internationalization of the Japanese diet after the 1850s and, again with greater force, after 1945.

To grasp the importance of the composition of diet, let us compare the Japanese diet in 1861 with selected years during the period 1874–1940. M. Umemura, N. Takamatsu, and S. Itoh (1983: 35) give the following percentage distribution of the main staples in 1861: rice, 47 percent; barley, 28 percent; assorted grains, 19 percent; and potatoes, 3 percent. Throughout most of Japan at the close of the preindustrial era, rice was the main source for caloric energy and for proteins.[5] The role of rice declined until the 1930s, when it increased briefly; since the war consumption of rice has been in decline, especially in urban areas.[6] Consider the calorie and protein intake arising from various sources (Mosk and Pak 1978).

	1874	1900	1920	1940
Calories				
Rice	59%	51%	53%	61%
Barley	7%	6%	4%	3%
Potatoes	6%	8%	8%	3%
Proteins				
Rice	34%	30%	31%	35%
Miso	8%	8%	8%	5%
Fish	22%	21%	26%	30%

The trend toward growing diversity in diet during the 1877–1920 period is apparent. However, during the interwar years there is a noticeable shift back toward rice, although the contribution of fish to overall protein intake seems to be on the rise during the interwar epoch. Perhaps even more dramatic in terms of a shift in the composition of diet antedating the American Occupation is the pronounced shift toward fruit as a source of vitamin A. The following figures give us important indicators of the impact of shifts in diet on the percentage composition of vitamin intake by food type for the years 1874 and 1940 (Mosk and Pak 1978).

	1874	1940
Vitamin A		
Fruit	32%	56%
Vegetables	21%	9%
Fish	34%	17%
Vitamin B1		
Rice	37%	39%
Vitamin B2		
Rice	26%	24%
Vegetables	16%	19%
Vitamin C		
Vegetables	76%	74%

Of course, these trends in diet may well be overshadowed by the dramatic increase in calcium and fat intake following World War II conditioned by a growing American influence in dietary matters. There was a definite shift toward coffee, cereal, toast, and juice for breakfast and toward lunches based around sandwiches and milk. Moreover, school cafeterias served milk at lunch. But as important as the postwar shift is, the extent of change during the prewar era should not be ignored.

If the linkage between human growth and diet were more accurately understood, it might be possible to focus on more specific indicators of food intake. Lacking this kind of detailed information, my approach is to explore the impact that two composite indexes (NUTI1 and NUTI2) and one more specific index (CALCFATI) have on human growth.

My approach to the construction of an index of public health and medicine is based on slightly different reasoning. The basic indexes underlying the overall composite variable are discussed in panel D of chart 2 and estimates for the 1900–1985 period appear in table 9. My idea was to use two criteria in constructing an index of the impact of public health and medicine on a reduction in the frequency and severity of disease: the level of inputs per capita (a quantity measure) and the efficacy of the inputs (a quality measure). Before the 1940s and the development of antibiotic drugs, the ability of doctors and hospital staffs to effectively treat disease was much more limited than it was after 1950. Hence my approach is to build up an index that recognizes that the ratio of doctors (and by inference other medical personnel like nurses

TABLE 9

Index of Public Health and Medicine, 1901-1985 with 1900-1904 = 100

Period	DOCPC[a]	DOCPCI[a]	IIDR[b]	IIDRI[b]	CPDR[c]	CPDRI[c]	PHMEI[d]
1901-1910	73.9	95.2	581.6	91.7	3.9	97.1	94.7
1911-1920	79.8	102.8	761.2	70.4	3.3	81.7	85.0
1921-1930	76.0	97.9	703.1	75.3	3.3	80.6	84.6
1931-1940	82.9	106.8	608.1	86.8	3.8	94.1	95.9
1941-1950	83.0	106.9	507.6	108.3	4.9	121.4	112.2
1951-1960	105.9	136.5	162.2	350.4	19.2	473.2	320.0
1961-1970	112.3	144.7	77.3	700.9	152.0	3,745.6	1,530.4
1971-1980	121.2	156.2	51.7	1,026.3	311.0	7,661.7	2,948.0
1981-1985	144.4	186.1	50.8	1,040.6	357.6	8,810.3	3,344.3

SOURCES: Japan Statistical Association 1987: table 2-33-b (pp. 242-245); Japan Statistical Association 1988: tables 21-9 (pp. 146-155) and 21-19 (pp. 178-179).

NOTES: [a]Levels and indexes for doctors per capita. Value for 1981-1985 is actually for 1981-1984. Death rate is per 100,000 population.

[b]Levels and indexes for the inverse of (one over) the death rate from tuberculosis, pneumonia, bronchitis, and enteritis. Value for 1981-1985 is actually for 1981-1984. Death rate is per 100,000 population.

[c]Levels and indexes for cases per death for cholera, dysentery, typhoid fever, and smallpox.

[d]Overall index for public health and medicine. Overall index computed by weighting each of the indexes—for doctors per capita, for the inverse of deaths per 100,000 population for tuberculosis, pneumonia, bronchitis, and enteritis, and for cases per death for cholera, dysentery, typhoid fever, and smallpox—by one-third and then adding.

and pharmacists) to population is important, but so is the effectiveness of the treatments offered. Of course, public health measures designed to stop the spread of microorganisms were also important. Hence I utilize two measures that implicitly take into account the impact of public health programs and the efficacy of therapy and inoculation offered by the medical community: the inverse of the death rate from four major infectious diseases—tuberculosis, pneumonia, bronchitis, and enteritis—which takes into account both incidence of infectious disease and treatment of it when it occurs; and cases per death for four other infectious diseases—cholera, dysentery, typhoid fever, and smallpox—which measures in inverted form the efficacy of the medical community in treating infectious disease when it occurs. I use inverses for death rates and the death-to-case ratio to construct an index that grows in value as efficacy improves and disease incidence declines. And I concentrate on infectious disease since I am mainly interested in children and young adults and it is infection that is the main source of illness for them.

What is striking about the variable measuring public health and medicine is its dramatic increase after the 1950s. This is testimony to the effectiveness of the antibiotic drugs in treating infection.

The last of the three proxy variables, that measuring child/youth labor input, is described in panel E of chart 2. Estimates of the composite variable and its two components for the 1900–1985 period appear in table 10. Two series exist, one covering the entire period and the other beginning after the war. There is little that needs to be said here about this variable. The trends in it are strongly affected by two factors: the growth of the educational system and the secular increase in the length of compulsory education within the system; and the shift from unpaid family work, exemplified by agriculture and forestry, to wage labor. Children were an important source of labor in small shops and on farms. Hence the shift away from this activity had a decided impact on the degree to which children were introduced into constant physical exertion from an early age. The variable is also an indirect indicator of the physical work levels to which mothers were accustomed, which presumably had some effect on birth weights of children.

In sum, in this section I have described the description of three composite variables for gross nutritional intake, for levels of and efficacy of public health and medicine, and for child/youth labor input. What impact these three variables had on net nutritional intake and hence on the levels and the tempo of human growth is the subject of the next section.

TABLE 10

Indexes of Child and Youth Labor Input, 1901-1985 with 1900-1904 = 100

Period	Males					Females				
	PPIM[a]	PWNE[b]	LFPR[c]	ICYI1[d]	ICYI2[e]	PPIM[a]	PWNE[b]	LFPR[c]	ICYI1[d]	ICYI2[e]
1901-1910	58.3	n.a.	58.7	97.5	n.e.	66.2	n.a.	52.3	97.7	n.e.
1911-1920	53.6	n.a.	52.5	88.5	n.e.	64.3	n.a.	45.2	89.8	n.e.
1921-1930	45.2	n.a.	47.1	77.0	n.e.	61.5	n.a.	40.6	83.5	n.e.
1931-1940	40.2	n.a.	42.8	69.1	n.e.	58.8	n.a.	38.0	79.0	n.e.
1941-1950	39.0	56.1	49.9	73.9	73.8	61.1	57.9	48.9	90.7	81.2
1951-1960	31.6	46.1	26.9	48.8	61.9	45.1	67.0	24.7	56.6	68.3
1961-1970	18.7	31.8	20.4	32.5	41.9	29.6	51.0	20.7	41.2	50.7
1971-1980	10.4	24.5	10.7	17.5	27.3	16.5	39.9	11.3	22.8	33.4
1981-1985	8.0	21.7	8.5	13.6	22.9	11.5	34.4	8.3	16.2	26.5

SOURCES: Umemura et al. 1988: tables 1, 2, 5, 6 (pp. 166-171, 196-201); Japan Statistical Association 1987: tables 3-4 and 3-8 (pp. 376-377, 390-395).

NOTES:
[a] Percentage of gainfully employed males/females in primary industry (agriculture and forestry).
[b] Percentage of gainfully employed males/females who are not employees (i.e., are self-employed or unpaid family workers).
[c] Labor force participation rate for males/females ages 10-19.
[d] Computed by calculating indexes (1900-1904 = 100) for PPIM and LFPR and weighting each index by .5.
[e] Computing by calculating indexes (1900-1904 = 100) for PPIM, PWNE, and LFPR, weighting each by one-third and adding.
n.a. = not available.
n.e. = not estimated.

LONG-TERM DETERMINANTS OF THE
ANTHROPOMETRIC MEASURES, 1901–1979

My goal here is to provide some measure of the relative success of the gross nutrition hypothesis in accounting for the secular trend in levels and tempo of human growth. It is important at the outset to be clear about the goals I have set for myself in the analysis that follows. What I show is that using the criterion of classical statistical theory one cannot reject the net nutritional hypothesis for Japan. I also show that given this provisional acceptance of the net nutritional hypothesis, changes in the demands placed on nutritional intake, rather than changes in the levels of gross nutritional intake, appear to be decisive to the secular trend in human growth and population quality.

Is such a modest aim satisfactory? There has been considerable debate on this point over the last two decades. There now seems to be some sort of consensus among scholars in the social sciences that an arbitrary distinction between the methods of science and those of the humanities, which was the pivotal notion of logical positivism, is not possible. Indeed, the prevailing view now seems to be that rhetoric and argument based on a notion of persuasiveness that goes beyond a mechanical recounting of statistical "tests" and "experiments" is the best we can aspire to in the social sciences (see Denton 1988; Klamer, Mc-Closkey, and Solow 1988; McCloskey 1994; Mirowski 1987). To accept this position is not to embrace an extreme version of relativism according to which there is no criterion by which we distinguish between competing hypotheses and viewpoints. But it does lead us to abandon the idea that any given statistical test is absolute and definitive. And it leads us to a pragmatic approach to the choice of statistical methods. In particular, it leads me to a strategy based on two criteria: use of forms for time series analysis, which allows me to estimate elasticities so I can differentiate between stronger and weaker effects, and use of cross-sectional and qualitative data as a supplement to time series analysis. Part II of this volume explores the latter kind of evidence.

In what follows I use log-log regressions to explore the impact of the three major factors underlying trends in net nutritional intake on the secular trend in human growth in Japan over the eight and a half decades between 1900 and 1985. The particular regression formats I employ are presented in chart 3. In the appendix to this chapter I provide a discussion of some of the technical reasons for using the particular functional forms underlying my results.

CHART 3

Basic Forms for Time Series Regression

For any variable $X(t)$ where t is time (i.e., the year in our case) let $LX(t)$ be the natural logarithm of $X(t)$ and let $DLX(t) = LX(t) - LX(t-1)$.

[A] Regressions with Levels of Anthropometric Measures as Dependent Variable

To illustrate these regressions (for levels of height, weight, chest girth, and the BMI for males and females separately at ages 6, 12, and 18) I give two examples, both involving heights for females age 6. Let $HF6$ = height for females, age 6; $HF12$ = height for females, age 12; $6NUT1$ = index for nutrition, #1 averaged over the 6 years beginning in year t; $6PHMEDI$ = index for public health and medicine averaged over the 6 years beginning in year t; and $6CYLFI$ = index of child/youth labor input for females averaged over the 6 years beginning in year t. Then the basic regression used to estimate the impact of nutrition, health, and medicine and child/youth labor input for age 6 is (ε is the error term)

$$DLH6 = a_0 + a_1 \, DL6NUT1(-6) + a_2 \, DL6PHMEDI(-6) + a_3 \, DL6CYLFI(-6) + \varepsilon$$

and the basic equation used to estimate the impact of the independent variables on height at age 12 is

$$DLH12 = b_0 + b_1 \, DLH6(-6) + b_2 \, DL6NUT1(-6) +$$
$$b_3 \, DL6PHMEDI(-6) + b_4 \, DL6CYLFI(-6) + \varepsilon$$

In regressions on DLH18, I use DLH12(-6) instead of DLH6(-6).

[B] Regressions with Six-Year Gains in Anthropometric Measures as Dependent Variable

To illustrate how I calculated the impact of nutrition, public health and medicine, and the child/youth labor input on six-year gains in height, weight, body mass index, and chest girth, I use gains in female height. I demonstrate my procedure with the gain in female height between ages 6 and 12, GHFA, as dependent variable. The basic equation I used is

$$DLGHFA = c_0 + c_1 \, DLH6 + c_2 \, DL6NUT1 + c_3 \, DL6PHMEDI + c_4 \, DL6CYLFI + \varepsilon$$

The results appear in table 11 (for height), table 12 (for weight), table 13 (for the BMI), and table 14. While there are differences between the findings for each of the measures taken separately, their consistency is striking. Indeed, as a practical matter, consistency is important to my pragmatic approach because I do not believe that any one "test" will ever be decisive. What is the main message to emerge from these results? It is twofold. First, the impact of gross nutritional intake as measured here is not great. For instance, results secured with the two broad composite variables for gross nutritional intake, NUTI1 and NUTI2, based on a wide variety of nutrients like calories, proteins, and vitamins, do not seem to have a consistent and positive impact on the levels and six-year gains in the anthropometric measures. And insofar as nutrition does seem to have a consistent statistically significant impact, it does so in terms of fat and calcium intake; it is the CALFATI

TABLE 11

Secular Determinants of Standing Height and Gain in Height, 1901-1979[a]

A. Estimates Based on Indexes Covering Entire Period

Height	1907-1979				1907-1940				1945-1979			
	H(-6)	NUT1	PHM1	CYL1	H(-6)	NUT1	PHM1	CYL1	H(-6)	NUT1	PHM1	CYL1
Males, 6	n.e.	-	-	-.03[3]	n.e.	-	-	-	n.e.	-	+.01[3]	-.03[2]
Females, 6	n.e.	-	-	-	n.e.	-	-	-	n.e.	-	+.01[3]	-.03[4]
Males, 12	-	-	-	-.06[2]	-	-	-.35[3]	-	-	-	+.02[1]	-.09[1]
Females, 12	-	-	+.02[2]	-.10[1]	-	-	-	-	-	-	+.02[2]	-.10[1]
Males, 18	-.13[2]	-	-	-	-.15[1]	-	-	-	-.15[1]	-	+.01[3]	-
Females, 18	-	-	-	-	-	-	-	-	-	-	-	-

Gains in Height	1901-1979				1901-1940				1945-1979			
	H	NUT1	PHM1	CYL1	H	NUT1	PHM1	CYL1	H	NUT1	PHM1	CYL1
Males, 6-12	-4.10[1]	-	-	-.29[2]	-5.26[1]	-	-	-1.26[1]	-3.44[1]	-	+.06[1]	-
Females, 6-12	-2.61[1]	-	-	-.28[1]	-	-	-	-1.50[1]	-2.71[1]	-	-	-
Males, 12-18	-6.79[1]	-	+.04[3]	-	-5.98[1]	-	-	-	-7.43[1]	-	-	-
Females, 12-18	-11.4[1]	-	-	-	-11.6[1]	-	-	-	-10.8[1]	-	-	-

NOTES: [a]Based on first differences of log-log regressions. See chart 2 and appendix to chapter 2 for a discussion. Abbreviations roughly follow those given in tables 8, 9, and 10. H = standing height; H(-6) for year t indicates standing height for persons 6 years or younger in the year $t-6$; NUT1 for year t is a six-year average for the years t through $t + 5$ based on the index of nutrition for the entire prewar and postwar period; PHMEI for year t is a six-year average for the years t through $t + 5$ for the index of public health and medicine; and CYL1 for year t is a six-year average for the years t through $t + 5$ for the index of child/youth labor input covering the entire prewar and postwar periods. In the regressions on levels, the three indexes are lagged 6 years. In the case of gains, they are not lagged. In the regressions on gains, H indicates the level of height in the year in which the gain begins. Values only reported if they are at least significant at the 15% level (two-tailed test).

TABLE 11 *continued*

B. Estimates Based on Nutrition and Child/Youth Labor Input Indexes Covering Postwar Period Only[b]

Height	1953-1979				1955-1979			
	H(-6)	CALFAI	PHMI	CYL2	H(-6)	NUT2	PHMI	CYL2
Males, 6	n.e.	-	$+.01^{4}$	-	n.e.	-	$+.01^{3}$	-
Females, 6	n.e.	-	-	-	n.e.	-	-	-
Males, 12	$+.64^{1}$	-	$+.01^{1}$	-	$+.68^{1}$	-	$+.01^{1}$	$-.04^{3}$
Females, 12	$+.21^{2}$	$+.04^{2}$	$+.004^{4}$	-	$+.18^{1}$	-	$+.01^{3}$	-
Males, 18	$-.27^{1}$	-	-	-	-	-	$+.01^{3}$	-
Females, 18	-	-	-	-	-	-	-	-

Gains in Height	1953-1979				1955-1979			
	H	CALFAI	PHMI	CYL2	H	NUT2	PHMI	CYL2
Males, 6-12	-	-	$+.04^{1}$	-	-1.91^{1}	-	$+.04^{1}$	$-.18^{3}$
Females, 6-12	-2.31^{1}	$+.16^{3}$	$+.02^{4}$	-	-6.39^{1}	-	-	-
Males, 12-18	-6.94^{1}	-	$+.04^{3}$	-	-6.39^{1}	$-.29^{4}$	$+.04^{2}$	-
Females, 12-18	-12.96^{1}	-	-	-	-11.07^{2}	-	-	-

NOTES: [b]See note to panel A for a discussion of the regression format and a set of basic definitions. Additional abbreviations: CALFAI = index of combined calcium and fat intake; CYL2 = index of child/youth labor input based on postwar data series not available for the prewar period (see tables 8 and 10).

Significance levels (two-tailed tests): 1, 1% level; 2, 5% level; 3, 10% level; 3, 15% level.

TABLE 12

Secular Determinants of Weight and Gain in Weight, 1901-1979[a]

A. Estimates Based on Indexes Covering Entire Period

Weight	1907-1979				1907-1940				1945-1979			
	W(-6)	NUT1	PHMI	CYL1	W(-6)	NUT1	PHMI	CYL1	W(-6)	NUT1	PHMI	CYL1
Males, 6	n.e.	-	-	-.06[3]	n.e.	-	-	-	-	-	-	-
Females, 6	n.e.	-	-	-	n.e.	-	-	-	-	-	-	-
Males, 12	+.34[1]	-	-	-.16[1]	-	-.66[4]	-	-1.07[3]	-	-	+.08[1]	-.17[1]
Females, 12	-	-	-	-.13[3]	-	-	-	-	-	-	+.08[1]	-.17[1]
Males, 18	-	-	+.02[3]	-	-	-	-	-	-	-	+.04[3]	-
Females, 18	-	-	-	-	-	-	-	-	-	-	-	-

Gains in Weight	1901-1979				1901-1940				1945-1979			
	W	NUT1	PHMI	CYL1	W	NUT1	PHMI	CYL1	W	NUT1	PHMI	CYL1
Males, 6-12	-1.59[1]	-	-	-.38[1]	-1.61[2]	-	-	-2.19[1]	+.27[3]	+.07[1]	-	-.89[1]
Females, 6-12	-1.66[1]	-	-	-.24[3]	-	-	-	-1.36[1]	-	+.04[3]	-	-.86[1]
Males, 12-18	-	-	+.06[2]	-	-.72[3]	-.72[3]	-.37[3]	-.49[3]	-	+.06[2]	-	-1.43[1]
Females, 12-18	-2.14[1]	-	-	-	-1.41[1]	-1.41[1]	-	-.77[1]	-	-	-	-1.31[3]

NOTES: [a]Based on first differences of log-log regressions. See chart 2 and appendix to chapter 2 for a discussion. Abbreviations follow those given in tables 8, 9, and 10. W = weight; W(-6) for year t indicates weight for persons 6 years or younger in the year $t-6$; NUT1 for year t is a six-year average for the years t through $t + 5$ based on the index of nutrition for the entire prewar and postwar period; PHMEI for year t is a six-year average for the years t through $t + 5$ for the index of public health and medicine; and CYL1 for year t is a six-year average for the years t through $t + 5$ for the index of child/youth labor input covering the entire prewar and postwar periods. In the regressions on levels, the three indexes are lagged 6 years. In the case of gains, they are not lagged. In the regressions on gains, W indicates the level of weight in the year in which the gain begins. Values reported only if they are at least significant at the 15% level (two-tailed test).

TABLE 12 continued

B. Estimates Based on Nutrition and Child/Youth Labor Input Indexes Covering Postwar Period Only[b]

Weight	1953-1979				1955-1979			
	W(-6)	CALFAI	PHMI	CYL2	W(-6)	NUT2	PHMI	CYL2
Males, 6	n.e.	-	-	-	n.e.	-	-	-
Females, 6	n.e.	-	-	-	n.e.	-	-	-
Males, 12	-	-	$+.04^2$	-	-	-	$+.04^2$	-
Females, 12	-	$+.15^2$	-	-	-	-	$+.03^3$	-
Males, 18	-	-	-	-	-	-	-	-
Females, 18	$+.87^2$	-	-	-	-	-	-	-

Gains in Weight	1953-1979				1955-1979			
	W	CALFAI	PHMI	CYL2	W	NUT2	PHMI	CYL2
Males, 6-12	$-.72^2$	$+.19^2$	$+.06^1$	-	$-.73^3$	-	$+.07^1$	-
Females, 6-12	$-.89^1$	$+.32^1$	$+.03^3$	-	-	-	$+.05^1$	-
Males, 12-18	-1.22^1	-	$+.06^2$	-	$-.94^2$	-	$+.07^1$	-
Females, 12-18	-	-	-	-	-	-	-	-

NOTES: [b]See note to panel A for a discussion of the regression format and a set of basic definitions. Additional abbreviations: CALFAI = index of combined calcium and fat intake; CYL2 = index of child/youth labor input based on postwar data series not available for the prewar period (see tables 8 and 10).

Significance levels (two-tailed test): 1, 1% level; 2, 5% level; 3, 10% level; 4, 15% level.

TABLE 13

Secular Determinants of Body Mass Index, 1901-1979[a]

(Estimates Based on Indexes Covering Entire Period)

BMI	1907-1979				1907-1940				1945-1979			
	B(-6)	NUT1	PH	CYL1	B(-6)	NUT1	PH	CYL1	B(-6)	NUT1	PH	CYL1
Males, 6	n.e.	-	-	-	n.e.	-	-	-	n.e.	-	-	-
Females, 6	n.e.	-	-	-	n.e.	-	-	$-.82^{3}$	n.e.	-	-	-
Males, 12	$+.31^{2}$	-	-	-	$+.39^{3}$	-	-	-	$-.46^{4}$	-	-	-
Females, 12	$-.42^{1}$	-	-	-	$-.50^{2}$	-	-	-	-	-	-	$+.06^{2}$
Males, 18	-	-	-	-	-	-	-	-	$-.21^{3}$	-	-	-
Females, 18	$+.17^{4}$	$-.16^{4}$	-	-	-	-	-	-	$+.35^{3}$	-	-	-

Gains in BMI	1901-1979				1901-1940				1945-1979			
	BMI	NUT1	PH	CYL1	BMI	NUT1	PH	CYL1	BMI	NUT1	PH	CYL1
Males, 6-12	-7.13^{1}	-	-	-	-8.69^{1}	-	-	-4.39^{2}	-6.35^{1}	-	-	-
Females, 6-12	-11.0^{1}	1.37^{4}	-	-	12.0^{1}	-	-	-	-5.94^{1}	-	-	$+.53^{3}$
Males., 12-18	-5.03^{1}	-	-	-	-4.60^{1}	-	-	-	-5.72^{1}	-	-	-
Females, 12-18	-3.09^{1}	-	-	-	-3.20^{1}	-2.49^{1}	-	-	-1.39^{1}	-	-	-

NOTES: [a]Based on first differences of log-log regressions. See chart 2 and appendix to chapter 2 for a discussion. Abbreviations roughly follow those given in tables 8, 9, and 10. B(-6) indicates BMI for persons six years younger in the year t - 6; NUT1 for year t is a six-year average for the years t through t + 5 based on the index of nutrition for the entire prewar and postwar period; PH for year t is a six-year average for the years t - 5 through t for the index of public health and medicine; and CYL1 for year t is a six-year average for the years t through t + 5 for the index of child/youth labor input covering the entire prewar and postwar periods. For the regressions on levels, the three indexes are lagged 6 years. In the case of gains they are not lagged. In the regressions on gains H indicates the level of height in the year in which the gain begins. Values reported only if they are at least significant at the 15% level (two-tailed test). See tables 11 and 12 for significance test indicators.

TABLE 14

Secular Determinants of Chest Girth, 1901-1979[a]

(Estimates Based on Indexes Covering Entire Period)

Chest Girth	1907-1979				1907-1940				1945-1979			
	C(-6)	NUT1	PH	CYL1	C(-6)	NUT1	PH	CYL1	C(-6)	NUT1	PH	CYL1
Males, 6	n.e.	-	-	-	n.e.	-	-	-	n.e.	-	-	-
Females, 6	n.e.	-.12[2]	-	-	n.e.	-	-	-	n.e.	-	-	-
Males, 12	-.22[4]	-	.01[4]	-	-	-	.11[4]	-	-	-	.01[1]	-
Females, 12	-	-	-	-	-	-	-	-	-	-	-	+.13[1]
Males, 18	-	-	-	-	-	-	-	-	-	-.12[3]	-	-
Females, 18	-	-	-	-	+.94[1]	-	-	-	-	-	-	-

Gains in Chest Girth	1901-1979				1901-1940				1945-1979			
	CG	NUT1	PH	CYL1	CG	NUT1	PH	CYL1	CG	NUT1	PH	CYL1
Males, 6-12	-3.32[1]	-	-	-	-6.49[1]	-	-	-.95[2]	-3.19[1]	-	.05[1]	-
Females, 6-12	-3.82[1]	-	-	-	-4.34[2]	-1.55[3]	-	-1.38[1]	-3.30[1]	-	.04[3]	-
Males, 12-18	-4.37[1]	-	-	-	-4.36[1]	-	-	-.69[3]	-3.24[2]	-.51[3]	-	-
Females, 12-18	-5.25[1]	-	-	-	-4.16[1]	-1.24[4]	-	-.74[4]	-	-	-	-

NOTES: [a]Based on first differences of log-log regressions. See chart 2 and appendix to chapter 2 for a discussion. Abbreviations follow those given in tables 8, 9, and 10. C(-6) indicates chest girth for persons six years younger in the year t - 6; NUT1 for year t is a six-year average for the years t through t + 5 based on the index of nutrition for the entire prewar and postwar period; PH for year t is a six-year average for the years t through t + 5 for the index of public health and medicine; and CYL1 for year t is a six-year average for the years t through t + 5 for the index of child/youth labor input covering the entire prewar and postwar periods. In the regressions on levels, the three indexes are lagged 6 years. In the case of gains, they are not lagged. In the regressions on gains, H indicates the level of height in the year in which the gain begins. Values reported only if they are at least significant at the 15% level (two-tailed test). See tables 11 and 12 for significance test indicators.

variable that gets the best results in this time series analysis. Thus one tentative conclusion that can be drawn from my analysis is that the shift toward dairy product consumption that was especially pronounced after World War II contributed to the secular trend in population quality. Second, secular trends in demands placed on nutritional intake seem to have dominated in the secular trend in population quality, especially during the prewar period. For instance, the elasticities on the proxy for child/youth labor input are especially large in the prewar period. After the war this index appears to be less important and the index of public health and medicine appears to be more important, although its estimated elasticity does not tend to be very large. These results accord with common sense. Before the war and the introduction of antibiotic drugs, the efficacy of public health and medicine was limited; and immediately after World War II, compulsory education was extended through to the end of middle school, which drastically reduced child/youth labor input.

Putting more social detail in these somewhat dry results is my aim in the second part of this book. But the results reported here are of interest and give support to the net nutritional hypothesis basic to this study.

SUMMARY AND CONCLUSIONS

During the first eight and a half decades of the twentieth century there was a dramatic improvement in population quality as measured by an assortment of anthropometric measures for schoolchildren. I have provided a wide range of data that document this secular trend for both males and females. It is the basis for my assertion at the beginning of this study that the person living in Japan today is in many ways a "giant" in comparison to a person living in Japan a century earlier.

I have also presented estimates for three composite variables—one for gross nutritional intake, one for public health and medicine, and one for child/youth labor input—whose trends over the last eight and a half decades strongly support the inference that net nutritional intake has dramatically improved in Japan. Finally, I have shown that the hypothesis that the trend in population quality is related to (caused by, I might assert) improvements in net nutritional intake cannot be rejected in time series analysis.

In short, I have a story to tell about the secular trend in population quality in Japan and evidence that supports that story. But the story is incomplete. In the real world demand matters as well; and in the give-

and-take of the real world, markets and nonmarket factors shape the way demand is voiced. Hence consideration of our second theme concerning demand brings us down from the realms of abstraction to what, I trust, the reader will find to be a more concrete story. This concrete story concerns the way in which particular social groups go about using the market and entitlements to secure the levels of population quality they demand. In particular, I argue that because the market played a major role in shaping the demand for population quality during the feudal period and because entitlements were balkanized during that era, a legacy was created which played a major role in shaping developments after the 1880s when the nation began industrialization. This is the story I turn to now.

APPENDIX

CHOICE OF FUNCTIONAL FORMS FOR REGRESSION ANALYSIS

As chart 3 shows, I use first differences of logarithms of my dependent and independent variables in carrying out the time series analysis described above. Why did I use first differences of logarithms? Because I wanted to cope with potential nonlinearities that might plague the analysis and also because I am interested in estimating elasticities. In addition, I found through the use of Dickey-Fuller tests that most of the variables had unit roots (see Maddala 1992: chap. 14; Pindyck and Rubinfeld 1991). The problem here is technical but can be described in a nontechnical manner as follows: do the variables analyzed by regression analysis satisfy the assumptions of classical regression theory? If the variables do not meet the requirements—and in time series analysis many variables do not meet the requirements because of unit roots and other problems—one must take some action to cope with potential difficulties in making inferences using the t-statistics and confidence bands on parameter estimates. In the case of the analysis discussed in section 2.4, I found that most of my variables failed to meet the stipulated requirements because they have unit roots. For this reason I used the standard procedure of taking first differences to cope with the attendant problems.

A more sophisticated approach was suggested to me: the use of vector autoregression (VAR) procedures. I decided to eschew this approach because it is most fruitfully used when one is interested in forecasting. In forecasting one is not very interested in individual parameter esti-

mates but rather in the relative reliability of the entire estimated structure of a set of equations for predicting one or several outcome variables. Forecasting population quality in terms of the anthropometric
measures is of no interest to me. As I stated earlier, I have opted for an
eclectic approach that does not rely on one particular set of tests and
one particular data set. Indeed, it is my view that the critics of logical
positivism in social science have established a valid methodological
point by demonstrating that any assertion of validity based on a technique, no matter how sophisticated, is really a rhetorical device and
therefore subject to debate. At any rate, this belief informs the pragmatic attitude toward statistical analysis employed throughout this
study.

The Market, Entitlements, and Human Growth

CHAPTER 3

The Tokugawa Legacy

Our focus now shifts to demand for population quality and in particular to the way demand was voiced through markets and social movements designed to assert the importance of the community over the market through the call for health-enhancing entitlements. I use the term "market," but to be more accurate I should use the phrase "the interplay of many markets" because my intention is to include labor and capital markets and the market for goods and services. As for the "health-enhancing entitlements," what I have in mind is the legal right to secure resources whether these are foodstuffs or public health and medical services. Buying and selling on markets provides one way of securing entitlements, but as I shall employ the term in what follows my concern is with entitlements determined outside of markets, that is, entitlements determined in the political and social arena.[1] In all societies the market and entitlements each play some role in determining the actual level of population quality for particular geographic regions and particular socioeconomic groups. What is peculiar to Japan—not unique to Japan, but especially salient in the Japanese case—is both an emphasis on market incentives and the balkanization of entitlements. Why this emphasis on the market and balkanization of entitlements simultaneously developed and the significance of this fact for the relationship between net nutrition and population quality is the focus of my analysis in Part II.

Because entitlement rights are determined at the level of community and government, we must necessarily consider political organization.

Japan's political organization has changed many times over the last four centuries but most dramatically three times: at the beginning of the Tokugawa period around 1600 when feudalism became well established and refined; at the time of the Meiji Restoration and during the Bakamatsu period when the feudal system was collapsing, that is, from around 1850 to 1880; and at the end of World War II and especially during the American Occupation. While the first two periods of political discontinuity may have been more wrenching for the Japanese populace as a whole, it is the third that probably marks the greatest change as far as the central government's role in and specific policies involving entitlements is concerned. For this reason an account of the entire modern period from the 1850s until the present would be a task requiring at least two volumes. And for this reason I focus on only one of the two subperiods of the epoch since 1850: namely, the period up to 1940. However, to understand how the balkanization of entitlements developed from the 1850s until 1940 we must consider the legacy of the feudal system for postfeudal Japan. In this chapter I briefly review the most important social and economic developments during the period from 1600 to the 1850s which shaped the subsequent balkanization of entitlements.

The chapter commences with a discussion of the key political features of early Tokugawa Japan that set in motion the economic and demographic evolution underlying the balkanization of health-enhancing entitlements in late (1720–1850) Tokugawa Japan. The remainder of the chapter concentrates on the role of the market in determining population quality for subpopulations and the role of entitlements in determining population quality for subpopulations.

THE BAKUHAN SYSTEM AND THE EXPANSION OF RICE CULTIVATION DURING THE EARLY TOKUGAWA PERIOD

Fresh from victory over a coalition of rival warlords at the battle of Sekigahara in 1600, Ieyasu Tokugawa and his allied warlords initiated a remarkable experiment in government designed to bring internal peace and harmony to a country that for centuries had been plagued with internecine warfare. The fruit of Tokugawa ingenuity was the *bakuhan* system, a dual system of government balancing limited devolution of power to fiefs (*han*) and limited centralization of authority in the hands of the Tokugawa family and its feudal allies and retainers (the *bakufu,* tent government, also known as the *shōgunate*). Directly linked to the setting up and refinement of the dual system of administration over the first half of the seventeenth century were two other policies that had momentous implications for the economic and demographic development of Japan: the partial demilitarization of the country through the forcible relocation of the sword-bearing warrior class, the *samurai,* from the villages into the administrative center of the fief to which they were attached, the castle town; and a policy of isolationism designed to ensure that the balance of power achieved through the dual administration system and demilitarization was not disturbed by the intrusion and meddling of foreign powers and foreign religions serving as path breakers for foreign ideologies and customs. At its root the Tokugawa system was based on pragmatic divide-and-rule principles designed to keep the country from once again fragmenting, falling into the pit of internal civil war, as it had done so many times in the past. Given the tradition of local military control, complete centralization was impossible but continuation of local military control was out of the question. Hence the bakuhan system emerged and evolved as an ingenious experiment designed to forge a compromise between strong local warlords and the preeminent warlord and his immediate allies bent on aggrandizing power for themselves. The most dramatic social and economic consequences of the forging of the bakuhan political compromise was a century of expansion in carrying capacity of the Japanese land area as exemplified by population increase and the amount of new land put under rice cultivation, coupled with rapid urbanization and the initial nourishing of a vibrant craft and protoindustrial economy in the region immediate to the Tokkaido route connecting the two most prominent bakufu cities, Edo (now Tokyo) and Osaka.[2]

Under the system of dual administration developed during the early Tokugawa period the central *bakufu* authority claimed about a quarter of the land for itself and allowed the remainder to be divided up into roughly 250 fiefs, selecting rulers (*daimyō*) for these fiefs from the ranks of the warlords who at the battle of Sekigahara were either victorious or defeated. The idea was to create a balance of power through the creation of a crazy quilt pattern of local administrative control in which no major coalition of powerful daimyō, bent on forming a military alliance to defeat the bakufu, could arise. Moreover, to bring the potentially restive lower-level military class, the samurai, under control, the central regime required that samurai loyal to a daimyō take up residence in the administrative capital of the fief assigned to the daimyō, thereby separating them from their economic and political base in the countryside where they engaged in farming. As a result, as the system became increasingly refined and formalized and samurai attempts at rebellion against loss of power were successfully quelled, the samurai ceased to perform active military duties, although they continued to be trained in military techniques, and they evolved into a bureaucratic class, responsible for managing the finances of the fiefs. In particular, they managed the collection of taxes in the form of volumes of rice from the villages under fief control. The samurai no longer collected rice, but they received rice stipends from the coffers of the fief to which they owed allegiance. And at least in the early Tokugawa period the amounts of these rice stipends depended almost exclusively on the rank of the samurai household within the pecking order established in the fief into which the retainer was born. However, many reform-minded fiefs found this type of ranking system overly rigid and abandoned it in favor of a competitive examination system in the latter Tokugawa (1720–1868) era.

Another act taken by the bakufu to weaken the centripetal force of local fief rule was the establishment of the *sankin kōtai* system, whereby daimyō were obliged to reside in luxurious domiciles within the confines of the bakufu's capital, Edo, on a regular schedule. This policy not only forced the daimyō to make extensive expenditures that were potentially ruinous for his fief's coffers but also subjected the local rulers to scrutiny by the central authority. Because the samurai were paid in rice and the daimyō were pressured to build magnificent estates in Edo and to maintain splendid castles within their castle towns, a market grew up for buying and selling rice and craft products and other goods demanded by the elite. Osaka, which was Japan's traditional merchant center, began to experience explosive growth in demand for its marketing services, and as a result the entire region contiguous to the road connecting Osaka to Edo began to flourish.

The isolationist policy was also aimed at reducing the ability of fiefs to establish independent political and military prowess. In 1639, the bakufu announced a policy of national seclusion (*sakoku*) whereby diplomatic relations with foreign powers other than China and Korea were effectively terminated. Contact with Europeans was limited to a tiny island, Dejima, in Nagasaki harbor, on which a small community of Dutch people was allowed to reside. The bakufu was determined to squelch any attempts by daimyō at forming alliances with foreign powers. As part of its sakoku policy the bakufu outlawed Christianity, which it viewed darkly as a vehicle for the foreign policy of powers like

Spain and Portugal, and to ensure that the peasantry did not adopt Christian practices, it forced all peasant households to register at the local Buddhist temple (the *shumon-aratame-chō* registers from which contemporary historians estimate vital rates for individual village populations are a by-product of this registration policy). One important consequence of isolationism was an end to the periodic outbreaks of epidemic diseases brought by ships from the Eurasian mainland. This worked to reduce mortality and hence stimulated population growth.

The direct consequence of these policies was the extensive development of riparian works bringing water for irrigation into lands that hitherto had been unsuitable for rice cultivation. Before the establishment of the bakuhan system military clashes between local villages over the diversion of water from one to the other and hence over rights to regular use of water at a stipulated time of the year had prevented the spread of rice cultivation to many reaches of the country. A classic example of how bakufu control removed this constraint is the development of the Kiso River basin in the area running across central Japan and into the ocean near Nagoya. After the bakufu brought in military retainers from a remote fief in southern Japan to stop warfare from breaking out at the confluence of three major rivers near the mouth of the Kiso River, the process of developing dikes and irrigation ditches along the river and hence of opening up new fields for rice cultivation (*shinden*) gathered momentum.[3] As the number of rice fields that could sustain households increased, so did the number of villages. As a result the demand for labor increased, and potential costs of having a large number of children were reduced since children who could not as adults be sustained on the lands farmed by the parental household could now find opportunities for branching off and starting their own families in newly established villages. Hence between 1600 and 1720 when this process of expansion in villages seems to have come to a halt at the national level (however, some regions declined and some expanded thereafter), the population expanded at a brisk pace. It is difficult to be precise about population totals in Tokugawa Japan because the samurai were not counted in any of the estimates and counts for the country as a whole are not available before 1721 (see Hanley and Yamamura 1977: 36 ff.), but the most widely accepted figures seem to put the 1600 population at around 18.5 million and the 1720 population at 26.1 million, yielding an annual increase in numbers of about 0.3 percent per annum. In any event, there does not seem to be doubt about the fact that the population expanded at a fairly vigorous pace between the battle of Sekigahara and 1720 and that it did so in large measure because of the expansion in rice cultivation.

There was an additional factor stimulating population growth during the early Tokugawa period: urbanization. As we have seen, urbanization was stimulated by two bakufu policies: the forcible removal of samurai from the countryside and hence the expansion of castle town populations; and the sankin kōtai policy that give a decided fillip to development of Osaka and Edo and to cities like Nagoya along the Tokkaido route between the two great metropolitan areas.[4]

In sum, the political solution to stopping internecine strife hammered out at the beginning of the seventeenth century—centered around dual administration, partial demilitarization, and isolationism—set in process economic and demographic developments that were not, and probably could not be, foreseen by the architects of the system. These developments would eventually cause the system to atrophy during the period known as late Tokugawa, that is, from 1720 until the Meiji Restoration in 1868 when the shōgunate disappeared. A detailed examination of this process of decay is not of interest here. Rather, our interest is in the development of health-enhancing entitlements and in those market forces that were intimately connected to the maintenance and promotion of the standard of living in terms of work capacities and capabilities during the late Tokugawa period. We now turn to these issues, beginning with the household and the market and then turning to community and entitlements.

HOUSEHOLD, LINEAGE, AND THE MARKET DURING THE LATE TOKUGAWA PERIOD

The household and the family system are at the center of the relationship between the demand and supply of labor embodying either high or low levels of population quality during the late Tokugawa period. The reasons are various but can be summarized in terms of the following considerations: for most individuals work meant work in an enterprise managed by one's family or by the household within which one resided; in cases in which household or family members worked outside the home it was the household head, not the individual, who tended to make contract arrangements concerning remuneration and conditions of work. That household and family are the center of the supply and demand for population quality in late Tokugawa has a crucial corollary: those agents demanding workers of a given level of population quality were also those investing in those workers and therefore supplying them. The incentive to enhance quality was built in—internalized, so to speak—since the family was simultaneously operating on both sides of the market.

During the late Tokugawa period both changes in the supply of labor conditioned by a decline in fertility among rural households and changes in demand for labor conditioned by an increase in the demand for labor that possessed a wide range of skills or could be quickly trained in the requisite skills encouraged an improvement in population quality. As a result, the role of the market in enhancing population quality was strengthened. This trend occurred in many sectors of the labor market dominated by family-run employment; and it also occurred in segments of the market in which bargaining between households by and large shaped labor contracts. Moreover, in the sectors of the labor market where households played a less pervasive role—for instance, in the market in which the samurai supplied labor services—deterioration in fief finances due to the burden of sankin kōtai also tended to encourage a favorable trend in population quality.

Since family and household play a crucial role in this argument we must say something about the stem family system that by the late Tokugawa period provided the basic set of rules within which families and households formed.[5] What is a stem family system? It is a system characterized by the following rules: (1) a spouse is brought into the family by one and only one offspring in each generation; (2) succession of the family headship falls on the offspring who has married within the family (or to the married couple); (3) inheritance, which is unequal, favors the single heir/successor; and (4) the family's organizational form passes through an alternating cycle of conjugal phases followed by a stem phase (in which the junior and senior conjugal units reside together) followed by a conjugal phase, and so forth. The transition from the stem to the conjugal phase is marked by the death of the last member of the senior couple; the transition from the conjugal to the stem phase is marked by the marriage of the heir.

That demography conditions the actual realization of a stem family system ideal is apparent from condition (4), for adult mortality and the timing of marriage play a critical role in lengthening or shortening the two phases of the alternating cycle. But demography conditions the operation of the family system in other ways. What if a family finds itself unable to biologically produce an heir, or produces an heir who is too young to assume the family headship at the same time the older couple wishes to "retire" and commence the stem phase of the cycle?[6] Had there not been a religious ideology buttressing the family system in Japan, families might have been content to calmly contemplate the possibility of their lines dying out. But because ancestor worship was deeply entrenched in Japan—in part because of Buddhism, in part because of long-standing domestic spiritual traditions—families had a strong aversion to the demise of lines (although lines did die out with some frequency among the poorer landless houses). Under the ideology of ancestor worship the Japanese family was conceived of as dynastic as well as stem, with the lineage extending through past and future generations through the worship of one's ancestors. For this reason securing a successor to assume the headship and attend to the family's relics and religious artifacts was a matter of paramount concern. And hence in securing heirs, families resorted to a variety of fictions. For instance, males not related to the family by blood ties were in-adopted as "sons." In this way family lines could be perpetuated even if biology failed. But in practice biology limited most families simply because relying on the market for in-adoption was risky—because supply might be limited at the time a household found itself searching for an heir; because in-adopted heirs might be less inclined to attend to the religious rites of the ancestor cult than children conditioned to do so from birth; and because an individual brought in to the house through in-adoption was not trained in how to operate the family's enterprise (if it had an enterprise to operate) from an early age.

Still families did employ in-adoption on occasion, even bypassing biologically produced putative heirs on occasion.[7] Why? Part of the problem was demographic and related to the timing of the transition from conjugal to stem phases of the cycle. But economics played a role as well. For it must be kept in mind that the actual working of the system was constrained by more than demographic probabilities. Most families were both demographic and economic units in the sense that they managed family enterprises like farms or craft production shops. If the biologically produced putative heir could not shoulder the task of managing the household economy in a highly competitive economy in which households vied with one another for land and rank, then the household head might be inclined to pass over his own son in favor of a more competent person unrelated by kinship.

Thus economics and demography constrained the family system in practice, if not in theory. But how actual adjustments were made very much depended on the social status, the class status, of the family. Here

I differentiate between two opposing class types: the peasant and the samurai warrior/bureaucrat. The reasons for differentiating in this way between two extreme types are both economic and ideological: economic, because in the peasant case household and enterprise overlapped and in the samurai case they did not; ideological, because in principle the code of behavior applicable to the representative samurai household was at variance with the code of behavior expected of the typical peasant household. I do not consider here merchant and craft families since they are less sharply opposed than the two types I focus on and since they were less numerous and have less bearing on my general argument.

In the case of the samurai the head of the house was invariably male as it was the male—not the female, who concentrated on the raising of a successor and domestic chores—who performed the tasks and maintained the rank and prestige of the house on which its fortunes rested. And among members of the bureaucrat/warrior class the income enjoyed by the house (*ie*) was a function of the rank that that male enjoyed with the domain's elite. K. Yamamura (1974) shows that there were marked differences in family support income allotments (measured in *koku* of rice allotted to the house) according to the position of the head. And while at the beginning of the Tokugawa period the position of the house was hereditary, financial pressures on fief coffers due to the burdens of sankin kōtai increasingly forced fiefs to become more efficient in allocation of scarce resources and therefore to seek out talented administrators among those retainers best qualified as signaled by performance in fief schools designed to train the retainer class. During the latter half of the Tokugawa period in many fiefs samurai were promoted upward and downward on the basis of ability and performance on examinations. As fiefs found their coffers depleted, many samurai, especially those of low rank, found their rice stipends inadequate for maintaining a large family in comparative luxury. And since their real incomes fluctuated with the price of rice on the Osaka market, in periods when rice prices declined the marginal retainers found themselves all the more squeezed. Hence is it surprising that family sizes among the samurai class declined and that many samurai found themselves without biologically produced sons or in need of a talented successor to replace a less competent putative biological heir? For these reasons the proportion of in-adopted family heads (*yoshi*) was quite high (Yamamura 1974). In short, there was a long slow secular trend among the retainer class toward reduction in fertility—with an attendant substitution of quality for quantity of children—and an increasing emphasis on ability and training in fief schools. Both factors tended to enhance population quality among members of this class. And this trend was directly related to changes in product markets like the market for rice and in financial markets where fiefs were forced to borrow from merchants to secure funds for their activities.

By contrast, among the peasant class the typical house served as a unit for organizing domestic activities and as a production unit in which all members regardless of sex, except the very young and the very old, were expected to perform the tasks of physical work on which the incomes of the house as a group depended. There was sexual division of labor—women devoted more time to domestic chores than did men and women were likely to engage in by-employments involving spinning and weaving (see Saito 1991)—but this division concerned differentiation within the categories of potential income-generating activities and did not draw a rigid line between those who restricted their activities exclusively to domestic work and those who engaged in market activities. Hence the de facto status of women depended on the actual activities they engaged in. If she possessed sufficient acumen and physical prowess, the mistress of a peasant ie might command greater authority than her husband. For this reason it is not surprising that the headship of the ie might devolve onto the mistress of the house or onto a daughter. Indeed Y. Hayami (1983) shows that male primogeniture, which is the form of inheritance usually associated with the stem family system in Japan, was not as common as was formerly believed. In some areas, for example, families employed ultimogeniture, and in some villages in the Northeast succession passed to the last born child regardless of sex (ane katoku). The point can now be made by contrasting peasant and samurai forms of the stem family system: because the peasant house tended to be both a kinship and a production unit, it displayed far more flexibility in gender differentiation and inheritance than did the samurai household.

Now, as with many samurai houses, but for somewhat different reasons, there was a secular trend among the peasant household sector toward an improvement in population quality during the late Tokugawa period. The reasons lie with both supply and demand, and since peasant houses tended to simultaneously operate on both sides of the market the conceptual distinction I make here between supply and demand is somewhat artificial. In any case, in many regions of Japan, especially in rice-producing areas, there was supply side drift as fertility declined, encouraging a substitution of quality for quantity. Why did this occur?

Since the occasional extant Buddhist temple register allows us to reconstruct with considerable accuracy for each village covered in the registers the age-specific pattern of fertility for women, we know that in
most cases interbirth intervals were long, overall marital fertility was
low, and life expectancies at age one were quite high. Findings on fertility and mortality levels for a sample of late Tokugawa villages appears
in table 15. As can be seen, fertility levels, although not low by modern
Japanese standards, are modest, and life expectancies tend to be moderate to high although variability is considerable. Moreover, since use
of the register data also allows us to reconstruct the sex ratios for sur-
viving offspring, we have evidence suggesting that infanticide (or
mabiki, literally winnowing out) was practiced in some villages, especially against girls (although there is some debate on whether infanticide was also systematically used against male issue in order to secure
sex balance among sibling sets). Since infanticide is an emotionally
charged issue, it is not surprising that the issue of why fertility reached
low levels in many districts of late Tokugawa Japan is one that gener-
ates heated controversy.[8] Without attempting to settle the issue here, I
will merely note that at least three positions have been staked out as to
the development of a low fertility/moderate mortality regime in late
Tokugawa Japan. T. Smith (1988: chap. 4) argues that the main factor
was the desire to limit competition over who would succeed to the
headship; S. Hanley and K. Yamamura (1977) argue that peasants mo-
tivated by a desire to emulate the conspicuous consumption prevailing
among urban dwellers limited their fertility so as to enhance their standard of living; and C. Mosk (1983) attributes the behavior to a desire to
maximize survivorship for each child not eliminated through infanticide, low fertility, and a long interbirth interval strategy to reduce the
number of infants requiring care at any one time. In any event, regardless of which explanation or combination of explanations is adopted, it
is reasonably clear that there was a shift from a high fertility regime
with frequent mortality crises and significant population growth during
the early Tokugawa period to a regime of moderate fertility and moderate mortality and virtually no population growth during the late Tokugawa period (see Kalland and Pedersen 1984). Not all the regions of
Japan experienced this sea change in demographic regime to the same
extent, but the general pattern is clear enough. And from the point of
view of this study the most important consequence of this secular
change was a substitution of quality for quantity among children,
as couples who restricted the overall size of their families were able to

TABLE 15

Fertility and Mortality in Selected Tokugawa Villages

Village	Period	MAFMF[a]	ALB[a]	TMFR[a]	LEM[a]	LEF[a]	IMR[a]
Asakusanaka	1717-1830	19.6	37.5	6.5	n.e.	n.e.	n.e.
Kabutoyama	1675-1780	18.3	39.8	6.1	n.e.	n.e.	n.e.
Kandoshinden	ca. 1800	21.6	39.2	7.3	33.2	31.6	n.e.
Minami Oji	1700-1899	17.9	n.e.	n.e.	37.1	38.4	n.e.
Nakahara	1700-1899	19.6	37.5	n.e.	46.1	50.5	165
Ogenji	1776-1875	n.e.	n.e.	n.e.	32.3	32.0	288
Shibuki	1826-1871	23.4	36.0	4.7	n.e.	n.e.	n.e.
Shimoshinjo	1828-1847	20.4	37.8	6.2	n.e.	n.e.	n.e.
Shimoyuda	1737-1870	15.6	29.9	2.6	n.e.	n.e.	n.e.
Toraiwa	ca. 1815	n.e.	n.e.	n.e.	36.8	36.5	229
Yokouchi	1700-1899	19.4	37.1	5.0	36.8	29.0	n.e.
Yufunezawa	1731-1765	20.2	39.3	5.0	n.e.	n.e.	n.e.

SOURCES: Morris and Smith 1985: 229-246; Mosk 1983: table 5.1; Saito 1993: table 1; Tomobe 1994: table 3.

NOTES: [a]MAFMF = mean age at first marriage for females; ALB = age of marriage at last recorded birth; TMFR = total marital fertility rate; LEM = life expectancy for males at age 1; LEF = life expectancy for females at age 1; IMR = infant mortality rate (males).
n.e. = not estimated or not available.

devote more resources, more foodstuffs, and more parental time to those whom they chose to raise and train in the techniques of agriculture.

The argument involving fertility concerns the supply of population quality, but secular drift in demand was also important. On the demand side two forces were active: technological changes in agriculture that increased the importance of skill formation, especially in rice cultivation; and the expansion of by-employment opportunities in rural districts. For instance, in rice production new seed varieties were developed especially in the Southwest where experimentation with fresh strains was encouraged by the relatively benevolent climate, making double-cropping of rice possible and hence the risk of total crop failure minuscule. Moreover, the market for soil-enhancing fertilizers like fish cakes and dried sardines and the like expanded as rice cultivation spread and increased fertilizer usage improved the inherent productivity of land. And new threshing machinery like the bamboo *semba-koki* was introduced. Most of these innovations did not save on labor as did the mechanization of agriculture in nineteenth- and twentieth-century North America. Rather, they increased the demand for skilled labor that could efficiently choose from among a growing variety of seed varieties, fertilizers, and harvesting techniques (Smith 1959, 1988). Care and knowledge increasingly commanded a premium as is attested by the growing prevalence of the practice of transplanting the rice plants to the fields in even rows to maximize absorption of the sun's energy. Moreover, since rice was taxed by fiefs but industrial crops were not, wealthier farmers began to divert their activities to a more diverse set of crops and even to craft production aimed at distant urban markets. The result was a growth in by-employments. Especially common in late Tokugawa was the practice of *dekasegi* (temporary) migration to a small silk filature, or the like, for a period of a year or two. Girls with a knowledge of silk raising and spinning techniques were especially in demand in regions like Gifu and Nagano prefectures in central Japan where the growing of mulberry leaves on which silk-generating cocoons were fed was common. In short, changes on the demand side of the labor market in late Tokugawa Japan increased the demand for population quality and hence proceeded in tandem with supply changes associated with the secular drift toward moderate fertility. Indeed, A. Hayami has argued that the increase in demand for female labor in by-employments actually worked to depress fertility by raising mean ages of first marriage in many late Tokugawa villages (see Mosk 1995a, 1995b). But as a perusal of table 15 makes clear, there was substantial regional variability in demographic regimes.

The issue of regional variability brings us to the question of the national and regional standard of living defined in terms of opulence. There has been a long-standing debate about the standard of living that was originally fueled by the question of how Japan was able to become the first country outside of the European cultural zone to successfully industrialize. One answer has been that her preindustrial income level was comparable to that of preindustrial Europe; as we have seen, recent research on late Tokugawa period village demography has suggested that fertility and mortality levels were moderate and some scholars have argued that this demographic regime was either a cause or a consequence (or both) of a reasonably good preindustrial level of income per capita. Backward projection of estimates of income per capita with real income growth rates has been commonly used as a method of coming up with figures for per capita income in the late Tokugawa and early Meiji period. Using this basic approach, S. Kuznets (1971) came up with a figure of $74 in 1965 U.S. dollars; K. Ohkawa, with $172 and $251 in 1970 U.S. dollars, the latter figure adjusted for purchasing power parity; and Y. Yasuba (1987), with $268 in 1970 U.S. dollars. By comparison, N. Crafts (1983) estimates the following for per capita income in 1970 U.S. dollars for various European countries on the verge of industrializing and having reasonably reliable data (date for estimate in parentheses): Great Britain (1760), $399; France (1830), $343; Germany (1850), $418; Italy (1860), $451; and Russia (1890), $276. With the exception of Russia, it can be seen that even the optimistic estimates for Japan like those of Yasuba (1986, 1987) put Japan below European countries in the decade before they began to industrialize. In my opinion this is not as surprising as one might think. For income per capita is an opulence measure of the standard of living, and my thesis is that if we were to measure Japan's income in the 1850s in terms of capabilities or in terms of population quality it would be considerable and that population quality, rather than the standard of living defined in opulence terms, is relevant for successful industrialization. But there is an additional consideration: the regional distribution of income. By establishing fief barriers to the diffusion of best practice technique, the bakuhan system kept pent up in the Southwest a variety of rice production techniques. Had the farmers in the remainder of the country known about these techniques, the national level of per capita income might have been considerably higher.

And of course climate played a role at the regional level both in terms of average levels of per capita output and in terms of the willingness to experiment with new seed varieties and fertilizers.

Hard data on the regional distribution of income are not easily obtained. But it is possible to come up with a few rough indicators. Hanley and Yamamura (1977) provide figures on rice output (in koku) and population for sixty-eight regions (*kuni*) of Japan around 1700 and 1870. They are skeptical about the reliability of these figures, especially for the latter year, but it is worth exploring what the figures reveal about variation in output per head. I ranked the kuni by the koku per head figures and then grouped the regions into seven classes, for each of which I calculated the average koku per head. (There are 10 kuni in each group except in that with the lowest level of income for which there are eight regions; the index given over on the far right column below was secured by setting the figure ca. 1700 at 100 and then calculating the 1870 figure relative to it. Kuni in both years are grouped on the basis of the 1700 figures.)

	Top	Second	Third	Fourth	Fifth	Sixth	Seventh
Koku per Head							
Ca. 1700	1.73	1.28	1.08	0.96	0.88	0.75	0.57
Ca. 1870	1.18	1.29	1.19	0.87	0.85	0.75	0.61
Index (1700 = 100)							
1870 value/							
1700 value	70.0	100.8	110.1	90.8	97.0	99.6	108.5

Substantial cross-sectional variation in agricultural productivity is evident from these figures. For what they are worth, the data also suggest that the differentials did not close substantially during the Tokugawa period. However, Hanley and Yamamura (1977) warn that the koku figures for the kuni were not revised frequently so it is possible that labor productivity increased more than we have estimated in the less productive regions. But there is no reason to believe convergence occurred during the late Tokugawa period.

By contrast, rapid convergence took place during the Meiji era (1868–1912). To see this consider the following figures for rice yield per hectare in Saga prefecture in southwestern Kyushu and for rice yield per 10 hectares in Tochigi prefecture, which lies in the northeast of the main island, Honshu, of Japan (Tsuchiya 1976: 60).

Year	Saga	Tochigi
1808–1812	454	231
1848–1852	429	225
1868–1892	420	250
1888–1892	430	296
1908–1912	447	340
1928–1932	453	392
1933–1937	442	412

How typical the Saga and Tochigi figures given here are for the regions in which they lie, Southwest and Northeast, respectively, is open to debate, but the picture of convergence drawn with these data accords with the descriptive literature on government and landlord promotion of the diffusion of best practice technique during the Meiji era.

That regional variation in climate, soil quality, and methods of farming generated substantial geographic variation in yields can also be demonstrated by examining the capacity of different regions to sustain population increase during the late Tokugawa period. Hence while for the country as a whole population growth over the 1721–1846 period was virtually nonexistent, there was considerable regional variation in growth rates. H. Kito (1983: 17 ff.) gives figures for the fourteen major regions of Japan over the period 1721–1746 and I have taken these data and calculated simple unweighted averages of population group rates in percentage terms for groups of these regions classified by an overall index of the warmth of weather (there are five regions in the first two groups and four in the third group, and the northeastern regions rank low in terms of the index while the southwestern districts rank high).

	Coldest	Middle	Warmest	Total
Warmth Index	96.0	118.2	128.3	114.1
1721–1846	−2.2	+4.5	+16.9	+3.0
Crisis Years	−16.6	−8.4	+2.3	−9.0
Normal Years	+14.4	+12.9	+14.6	+12.1

The term "crisis years" refers to the periods of food shortages, especially to the major famine era, including the Kyoho (1732–1733), Tenmei (1783–1786), and Tempo (1836–1838) famines. Note that regional variation in population growth rates is substantial during crisis periods but not during normal years. Thus whatever is checking population growth is related to a realized or potential threat of food shortage. That this appears to be the case makes one a bit skeptical of explanations for low fertility that emphasize a positive linkage between the level of income per capita and the willingness to restrict fertility.

In sum, the family lay at the center of the adjustment of population quality to changes in labor markets: it was the unit that determined supply, and for the great mass of the people who were peasant producers it was also the unit determining demand for skills, work capacity, and capabilities. In the regions of the country where technological progress in rice farming or the expansion in the demand for by-employments was especially vigorous, there seems to have been a secular trend toward improvement in population quality. This improvement was driven by a decline in fertility and an increase in the amount and quality of training accorded offspring. That the market was as effective as it was in enhancing population quality is one reason why entitlements played a more circumscribed role. However, as we shall now see, entitlements did play a role, especially in the fiefs where food crises occurred throughout the late Tokugawa period.

COMMUNITY AND ENTITLEMENTS
DURING THE LATE TOKUGAWA PERIOD

As pervasive as the market was in promoting improvements in population quality during the late Tokugawa period, entitlements supplemented the market. During this time the most important entitlements were those that provided insurance of sufficient food to eat in times of dearth for individual households and villages. Because entitlement insurance in the late Tokugawa period tended to be spread out among fairly small numbers of households—at the level of the village the number of households was of course small, but even at the level of the fief, which was the largest unit involved, population size typically ran into the low 100,000s—I refer to the "balkanization of entitlements." By this term I mean to highlight the local nature of the guarantee and distribution of entitlements in feudal Japan.

The most salient of the entitlements over foodstuffs were those offered by the fief. Villages paid taxes in the form of rice. What did they get in return? Certainly the fief provided protection and quelled local uprisings that might spill over from village to village. They also organized the creation of ditches, dams, and other riparian works. But most important, fiefs provided rice and other foodstuffs in times of dearth. S. Vlastos (1986) characterizes many of the three thousand or so peasant

rebellions (*ikki*) that broke out during the Tokugawa period as political
movements on the part of the peasantry, serving to remind the fief that
it had an obligation to extend "benevolence" down to the villages when
harvests were poor. It must be kept in mind that the peasants had lever-
age in their relationship with the fief and that organizing ikki was one
way of sending a signal to the daimyō—households could steal across
fief boundaries and attempt to settle in the villages of neighboring fiefs
if local conditions were unbearable, thereby reducing the tax-producing
capacity of their home fief—or of putting pressure on the daimyō since
the bakufu did from time to time remove from office daimyō whose sub-
jects were becoming too restive. Typically the villages wanted either
lower taxes or outright supplies of grains. Vlastos (1986) argues that
during the early Tokugawa period ikki were mainly organized by village
headmen, but that by the later Tokugawa period village landowners
often organized rebellions to demand a redress of grievances against the
headman or other groups of landowners. For instance, he argues that
one reason ikki tended to break out in silk-raising areas toward the
close of the Tokugawa period was that silk-producing families often devoted little land to rice cultivation and hence they were unusually adversely affected when rice prices rose in times of dearth since they had
to purchase most of the rice they consumed. In short, groups of peasants did not want to abolish or negate the operation of the market with
entitlements, but they did want the fief to provide them with insurance
in the form of a backup in times of dearth, especially in times when demand outstripped supply on the rice market.[9]

The fief was not the only organization providing entitlement insurance to peasant households. At the village level poorer households often
sought the protective benevolence of more economically prosperous
households and formed *dozoku* units—that is, extended households
often bound together in fictive kinship terms—within which the economically inferior parts supplied labor services in exchange for access
to land for production and for foodstuffs in time of need. C. Nakane
(1967) argues that these dozoku, which have been extensively studied
by anthropologists, are usually not groups tied together by true bonds
of kinship but rather are economic organizations that in effect exchange
labor services for use of land and insurance services. Moreover, even in
areas of Japan where dozoku were absent, landlord-tenant relations
that were well developed by the later Tokugawa period, especially in the
commercially oriented region contiguous to Osaka and to the Tokkaido
route leading from Osaka to Edo, often involved insurance entitle-

ments. For instance, A. Waswo (1977: 29 ff.) notes that landlords often assumed fictive roles as "parents" of tenants, or as "grandparents" (of tenants of tenants), and in exchange for the rent that they extracted from their tenants were expected to provide rent reductions or even outright grants of food in times of dearth.

THE LEGACY: THE STRENGTH OF THE MARKET AND THE BALKANIZATION OF ENTITLEMENTS

For two and a half centuries, from the early seventeenth century until the Meiji Restoration in 1868, the bakuhan system guaranteed a high level of internal peace and stability to Japanese peasants. As a result rice cultivation and population spread rapidly during the first century or so of Tokugawa rule. At this point, around 1720, population growth at the national level seems to have stopped, although at the regional level the pattern is more complex and variegated. Why population growth stopped is a matter of debate, but there seems to be considerable evidence that it slowed down because fertility declined, perhaps in response to a growing scarcity of land that could be readily brought into cultivation, perhaps because households found themselves increasingly pressed and therefore concerned about economizing on scarce resources by maximizing the survivorship of children through the lengthening of interbirth intervals, and so forth. In any event the decline in fertility tended to improve population quality on the supply side. And on the demand side the same households who by limiting supply enhanced quality also demanded greater work capacity and capabilities from their offspring. For agriculture grew increasingly sophisticated in terms of labor requirements. And the expanding market for by-employments demanded flexible and ready skill development. For these reasons market forces played a central role in the development of population quality and the standard of living defined in terms of capabilities during the late Tokugawa period.

Entitlements that supplemented the market were less pivotal but were still important. The most important of these were insurance entitlements over foodstuffs and were provided at a fairly local level by fief governments. And at an even more local level dozoku and landlords provided insurance entitlements in time of dearth. In short, balkanization of entitlements was a fact of life under the system of bakuhan dual administration.

Thus strong market orientation and balkanized entitlements were the legacy that late Tokugawa Japan offered to a country that, after the Meiji Restoration, rapidly moved down the path of industrialization. The profound imprint of this legacy on the relationship between economic development and population quality during an era when, as Western technology poured into the country, labor markets only gradually took on a complexion differing from that characteristic of the late Tokugawa period is the subject of the next chapter.

CHAPTER 4

Population Quality in an Era of Balanced Economic Growth, 1880–1920

During the first half century of industrialization following the Meiji Restoration, economic growth was balanced in the sense that levels and growth rates in wages and labor productivity were roughly equal in agriculture and other sectors employing traditional Japanese technology and in the new industries utilizing a combination of imported Western techniques and traditional Japanese methods. As a result, labor moved smoothly back and forth between the two sectors of the economy. Discontinuity introduced with Western machinery and production methods was largely mitigated by the continuity of labor recruitment practices inherited from the late Tokugawa period when by-employments flourished in many rural districts and young women went out for several years of dekasegi work; indeed, it was females who predominated in the labor forces of the new textile plants using British-style ring spinning devices and mules, just as women had predominated in dekasegi migration streams during the late Tokugawa period. But the nature of contracts and work conditions changed as the large impersonal factory with constant turnover and a nebulous connection to the households supplying short-term labor to it supplanted the smaller operations of the Tokugawa period. Employers in large concerns did not have the incentives to preserve the health of their workers. For this reason, as the locus of employment gradually shifted away from the family-managed enterprise, there was a slow but inexorable breakdown in the market institutions that had favored an enhancement of population quality during the late Tokugawa period.

Moreover, there was a breakdown in the balkanized system of entitlements developed during the late Tokugawa period. In part, this occurred because of administrative changes as the fief governments and the villages under them were abolished or restructured and a system of prefectural, city (*shi*), and rural county (*gun*) governments—with town (*machi*) and village (*mura*) governments underneath county administration—was brought in to replace the feudal structure. Political decision making became more centralized, and the structure of balkanized insurance entitlements was significantly weakened in concept as well as in practice. Lacking a domestic model for how to reorganize entitlements, the central government shied away from demand for population quality policies, its innovations taking on a supply side bias favoring the importation of Western medical and public health technology. The breakdown in entitlements also resulted from the reshaping of landlord-tenant relations that had been a cornerstone of balkanized entitlements during the late Tokugawa period. Productivity gain in agriculture associated with a diffusion and refinement of best-practice Tokugawa technique diminished variance in harvest yields, thereby largely eliminating the dependence of tenants on landlords; and at the same time, to the degree that the return on agricultural investments fell relative to the return on investments in industry, innovating landlords began to lose interest in agricultural activities, thereby surrendering their roles as spearheads of improvement in rural Japan.

Three factors kept this breakdown in health-enhancing market and entitlement institutions from generating political and social unrest sufficient to force the government and large employers to drastically revamp their practices and establish a set of market- and entitlement-based substitutes for those that had previously served households and communities. The first factor was the legacy of the Tokugawa period and the implicit assumption that employers would naturally look after the health of their employees. It was widely believed in many circles that government coercion was likely to be unfairly enforced and might even be counterproductive. Governmental responsibility—insofar as it extended beyond its own employees—was best left to taking care of food crises and the like. And because of improvements in agricultural productivity, these crises were infrequent. The second factor was the nature of labor markets in the nascent large industrial sector, in particular, the high turnover and the predominance of female workers employed on short-term contracts. Unlike a farm household that invested in its own family members with the expectation of securing years of productive service, or invested in its members because when they left the household for marriage a long-lasting bond might be established between the household losing the child and the household gaining the child, the owners and managers of large spinning and weaving companies lacked a clear imperative to actively protect the health of their employees. Moreover, the mills tended to recruit workers from the ranks of small-scale tenant households, the least healthy and physically robust segment of the rural population, which further discouraged active investment in worker health. Village entrepreneurs who operated small spinning filatures and risked losing a good reputation if a girl's health was ruined might have a strong incentive to develop aggressive health-enhancing practices. But for the large spinning and weaving companies, whose female workers moved in and out of the labor force, the situation was different. And since the girls themselves did not expect to stay with the industry for a sustained length of time, they lacked the drive to actively organize unions or other employee-based organizations aimed at voicing a demand for better conditions. The third factor was that most employment continued to be located within the world of household-managed enterprises like farms and small shops, although there was a clear drift away from this sector over time. To some extent the problem was restricted to a relatively small portion of the population during the early Meiji period. As long as those experiencing the dirt and noise of mill work, the inadequate ventilation, the related outbreaks of pneumonia and bronchitis, and the possibility of succumbing to the ravages of tuberculosis were almost invisible, society at large took scant interest.

In short, those factors that had sustained and encouraged an improvement in population quality during the late Tokugawa period now acted as an impediment to further improvements. As the breakdown in entitlements occurred, differentials between groups favored and not favored by the market tended to widen. Unfortunately, systematic evidence on this breakdown and divergence in population quality is at best fragmentary. Therefore, in sifting through the various strands of evidence I will range widely. My discussion begins with a description of balanced economic growth during the Meiji Restoration, turns to the technological bias in government policy, moves on to the issue of the health and physical well-being of female factory workers, and closes with an analysis of regional differentials in population quality. To fully understand the significance of the evidence assembled here, it is necessary to compare the findings in this chapter with those in the next chapter, which deals with the interwar period when a new set of market and entitlement institutions began to emerge and take shape.

THE ECONOMICS AND DEMOGRAPHY
OF BALANCED GROWTH

During the period from the Meiji Restoration until World War I productivity in agriculture grew, as did productivity in the nonagricultural sector, and while growth in manufacturing outstripped growth in farming and fishing, differentials in growth were not large. At the time of the Meiji Restoration a common labor pool moved back and forth on a regular basis between farming and by-employment, including small-scale light industry, and the forces of supply and demand tended to equalize earnings and marginal labor productivity in the agricultural and nonagricultural sectors. Because differential labor productivity growth after the 1870s did not unduly favor the nonagricultural sector and because the vast majority of the newly created industrial jobs were in textiles and food processing, which made extensive use of short-term female labor that did not require long periods of training, employers had little incentive to offer premium wages to keep workers who by dint of job experience were especially skilled. As a result, marginal productivity of labor in agriculture set a wage floor for manufacturing and there was almost no divergence in earnings between manufacturing and agriculture.[1]

One of the reasons growth was balanced was because productivity of farmworkers increased as population quality improved. Consider the figures in panel A of table 16. While the number of farm families remained virtually unchanged throughout the 1880–1920 period, the number of workers per farm family gradually declined, mainly because the number of female workers per farm household was dropping. For example, the ratio of female farm household workers to male farm household workers, always less than 1.00 throughout the period, declines somewhat over the four decades. Now one reason why farm households could increasingly afford to release workers for dekasegi employment in factories is apparent from the figures on workdays per worker and on labor productivity: both days worked and productivity per workday go up. As each worker became potentially more efficient, the household was better able to "lend" some of its members to industry without much sacrifice in output. With the steady improvement in

TABLE 16

Selected Economic, Social, and Demographic Characteristics of Japan, 1881-1920[a]

A. Agriculture

Period	Workers per House in Farm Households				Properties of Arable Land (1934-1936 = 100)[b]				
	No. (000s)	Male	Female	Female/male	Area	Price	Rent (Re)	Productivity (Pr)	Pr/Re
1881-1890	5,472	1.42	1.23	.87	n.e.	n.e.	n.e.	n.e.	n.e.
1890-1900	5,467	1.40	1.20	.85	n.e.	n.e.	n.e.	n.e.	n.e.
1901-1910	5,500	1.39	1.17	.85	87.7	85.9	91.3	53.7	85.9
1911-1920	5,539	1.37	1.15	.84	95.3	92.8	90.2	115.7	92.8

Period	Labor in Workdays (WD) and Real Production per Workday (LP)[c]			Wages or Fertilizer Prices Relative to Land Rent[d]			Fertilizer or Machinery Input per WD[e]		Terms of Trade[f]
	WD	Male WD	LP	Wages	NITF	PHOF	Fertilizer	Machinery	Agric./Manuf.
1881-1890	113	131	100	n.e.	n.e.	n.e.	1.6	31.6	.75
1890-1900	131	150	119	n.e.	n.e.	n.e.	1.9	31.7	.77
1901-1910	139	160	137	91.3	252.7	163.6	4.6	36.1	.74
1911-1920	163	187	159	90.2	195.4	114.3	6.8	42.4	.75

NOTES: [a]Some series (in all three panels) are for 1880-1889, 1890-1899, etc.

[b]Indexes for rents, prices, and productivity of land are based on nominal figures and pr/re indicates the relative level of nominal land productivity (value added net of depreciation on capital assets) relative to an index of nominal land rent.

[c]"Male WD" refers to male equivalent workdays. Index for LP is with 1880 = 100. Here 1881-1890 figure is for 1880, 1890-1900 figure is for 1900, 1901-1910 figure is for 1910, and 1911-1920 figure is for 1920.

[d]Ratios of indexes (all indexes having 1934-1936 = 100) with land rent index as the denominator. NITF = nitrogen and PHOF = phosphate fertilizer.

[e]Fertilizer and machinery inputs in 1934-1936 prices and per 100 workdays.

[f]Price index of agricultural goods divided by price index for manufactured goods (1934-1936 = 100).

TABLE 16 *continued*

B. Income and Consumption per Capita and Government Spending on Social Security and Welfare

Period	Income per Capita and Consumption per Capita, Total and by Type[g]							Government Expenditure[h]
	GDPPC	CONPC	FOODPC	HOUPC	MEDPCPC	EDRECPC		
1881–1890	108.7	97.1	63.1	11.4	3.2	3.9		1.5
1891–1900	130.7	120.3	73.5	12.3	4.0	7.4		1.9
1901–1910	148.1	123.4	73.7	15.0	3.4	6.9		5.7
1911–1920	176.8	141.6	86.0	13.9	4.3	8.9		5.0

C. Relative Sectoral Wage Levels, Structure of Production, and Female Labor Input in Manufacturing

Period	Female/Male Wage Ratio		Agriculture/Manufacturing Wage Ratio		Urbanization[i]	PPI[j]	Hours, female[k]
	Agriculture	Manufacturing	Males	Females			
1881–1890	.65	.48	.93	1.23	n.e.	n.e.	n.e.
1891–1900	.72	.52	1.12	1.56	7.6	n.e.	76.5
1901–1910	.79	.46	.96	1.66	9.7	61.2	79.8
1911–1920	.75	.48	.95	1.47	10.6	57.7	79.8

SOURCES: Various tables from Hayami 1975; Japan Statistical Association 1987; Ohkawa and Shinohara 1979; Umemura et al. 1966; Umemura et al. 1988.

NOTES:

[g]All figures in 1934-1936 prices. GDPPC = gross domestic product per capita; CONPC = total consumption per capita; FOODPC = food consumption per capita; HOUPC = housing consumption per capita; MEDPCPC = expenditure on medicine and personal care per capita; EDRECPC = expenditure on education and recreation per capita.

[h]Percentage of central government expenditure for social security (including public health and medicine).

[i]Percentage of the population living in the six big cities of Tokyo, Yokohama, Nagoya, Kyoto, Osaka, and Kobe. Figure for 1901-1910 is average for 1903 and 1908, figure for 1911-1920 is average for 1913, 1918 and 1920; etc.

[j]PPI = percentage of the gainfully employed population (both sexes) in primary industry (agriculture and forestry). Figure for 1901-1910 is actually for 1906-1910.

[k]Percentage of hours supplied in cotton spinning by female workers; 1891-1900 figure is actually for 1895-1899.

population quality and with the lengthening in number of years mandated under the compulsory education requirement, the physical and mental capacity of those who entered into agricultural pursuits was enhanced. Moreover, as can be seen from the figures on fertilizer and machinery inputs, the worker also benefited from more fixed (machinery) and variable (fertilizer) capital. In this context it is useful to keep in mind that increasing the application of fertilizer not only improves the inherent quality of the soil but also, by curtailing the amount of time devoted to weeding, reduces labor requirements (see Brandt 1993). Indeed as farm children, who often did the weeding, were withdrawn from regular work to attend school, presumably the demand for fertilizers increased. In short, the secular improvement in population quality commenced during the late Tokugawa period and continued into the Meiji period as traditional best-practice techniques (the so-called *rōnō gijutsu*, technology of veteran farmers) diffused throughout the country under the active promotion of the government. This diffusion raised per capita agricultural output, and in its wake per capita gross nutrition also improved. And the effect of these trends was to free up workers for manufacturing, thereby encouraging the continuance of a balanced growth process begun during the late Tokugawa period.

Hence under conditions of balanced growth the agricultural sector and the legacy of techniques built up in the more productive fiefs made a major contribution to economic growth both in terms of generating output growth and in terms of elastically freeing up labor for manufacturing activities. Some indicators of the quickening pace of growth in nonagricultural activities and of its implications for urbanization and income growth can be seen in panel B of table 16. Note that the proportion of the population living in the six big cities where industrial production tended to be concentrated—five of the six big cities were either in the Tokyo (formerly Edo) or Osaka region or lay along the Tokkaido where by-employments had most vigorously flourished during the late Tokugawa period—grows rapidly and the proportion in primary industry declines. With the shift out of agriculture into the manufacturing sector, which enjoyed labor productivity growth rates somewhat in excess of those experienced by agriculture, real income and consumption per capita grew slowly but steadily (see panel B of table 16). It is important to keep in mind that this growth in manufacturing production during the 1880–1920 era was mainly restricted to light industry and that in light industry female workers predominated. Consider the figures on percentage of hours in cotton spinning supplied by females. As can be seen from panel B over 75 percent of the hours were supplied by females and this proportion actually grew during the period 1890–1900. Moreover, wages for the factory girls actually fell short of wages paid females in farming; being a seasonal activity, farming could not provide the regularity of work offered by many textile enterprises.[2]

What about male workers for whom—as panel B of table 16 demonstrates—manufacturing wages usually exceeded agricultural wages, albeit not by a large margin? Why does the wage pattern vary between the sexes? The critical differences were the skill levels required and the nature of the labor contract. Male employment was largely concentrated in government-managed military arsenals, mines, and a few heavy industrial facilities (some of which were later sold off to the private sector) and in a relatively small number of privately operated heavy industrial concerns that manufactured ships and the like. During the Meiji period when labor capable of employing the techniques necessary for using Western machinery was in short supply, a typical pattern was for these factories to contract out work with labor bosses who brought in with them to the shop floor their crews of subordinate workers. The labor boss trained these subordinate workers and allocated wages to them from the overall payment negotiated between him (it was always a male) and the firm. In fact, the labor boss system was a carryover from the Tokugawa period and is one more example of balkanized entitlement insurance as the labor boss "took care" of the basic subsistence needs of his subordinates in good times or bad. Fearing that the labor bosses could and would use their oligopoly position to extract excessively high rents out of the enterprise, corporations attempted to establish direct control over their workers, replacing the free-floating labor boss with the co-opted labor boss or internally promoted foreman whose fate was more closely linked to the company. Increasingly, as heavy industry grew and established direct managerial control over the shop floor, special status within companies was accorded professional workers graduating from the higher educational system and co-opted labor bosses. These favored workers were called *shain* (literally company workers), and as an elite they earned high wages, wages the enterprises tied to seniority and age in order to reduce turnover. Corporate paternalism in Japan initially developed around this elite and was associated with wage payments that, in comparison with agriculture, were generous.[3]

The vast majority of male industrial workers, however, were not given shain status. Until the slowdown in growth of the nascent heavy industrial sector, coupled with an expansion in the number of graduates from industrially oriented vocational schools during the 1920s, spelled the end of excess demand conditions for skilled labor, most highly trained blue-collar workers moved about with considerable frequency. Some moved with labor bosses and some moved on their own initiative. Consider, for instance, figures for the Shibaura and Ishikawajima engineering works in 1902 (Gordon 1985: 35):

	Shibaura	Ishikawajima
0–6 Months	24.7%	12.7%
7 Months to 1 Year	14%	10.1%
6 Years or More	16.8%	18.3%

Not only did blue-collar workers leave plants with frequency, employers often complained of poor work habits and lax work discipline among their ranks and as a result often resorted to firing these workers who were denied shain status and were known as *koin*, which implies hired worker/outsider status. For these workers wage levels were not especially generous, although the more skilled they were, the greater the premium. At the time the typical heavy industrial enterprise saw no incentive in offering a generous seniority-based wage package as a vehicle for eliciting loyalty and effort.

As it had during the late Tokugawa period, female dekasegi labor continued to be the backbone of the industrial labor force. In other words, despite the great advances being made in importing and adapting American and European technology in Japan, continuity in the labor market remained strong in the sense that high turnover and the short-term labor contract prevailed. What was new was the fact that large plants operating with central power sources like steam engines or with electricity began to supplant the small rural workshop of late Tokugawa. And what was new within the large confines of textile plants or in the depths of coal mines was the danger of industrial contamination and exposure to airborne infections that had not been prevalent in the small shops of the late Tokugawa period. In short, as the large enterprise employing the bulk of its workers on a short-term basis increased its share of the total labor force, the proportion of employers with an active incentive to protect the health and physical well-being of their workers declined. And at the same time the health of the industrial work environment deteriorated. Slowly but steadily the health-enhancing nature of the Japanese market was breaking down. And this was taking place in an environment in which the government was reluctant to intervene for reasons to which we shall now turn.

COMMUNITY AND GOVERNMENT POLICY

During the Tokugawa period, the strength of the market and the balkanization of entitlements under the dual administrative system of bakuhan rule kept the nominal bakufu government of Japan out of entitlement programs aimed at the mass of its populace. For this reason the new Meiji government was reluctant to pursue an activist entitlement policy. But as Japan opened herself up to trade and to contact with the West, it became apparent to the former samurai who, as the educated elite, assumed responsibility for guiding the country along the path to industrialization that foreign models could be profitably and efficiently studied and adapted in many areas outside of that involving industrial machinery. As a rational borrower Japan could pick and choose the countries it wanted to imitate depending on the type of institution involved: thus the Meiji government designed its new police system along French lines, its military along German lines, its higher academic system along German lines, and so forth. But two factors constrained the government in its eagerness to follow Western guidelines: financial and resource limitations and potential resistance to social engineering, namely, the extent to which Tokugawa institutions were so deeply rooted that new institutions were unlikely to be accepted, or accepted only at the expense of social unrest. Financial constraints and the Tokugawa legacy of balkanized entitlements and reliance on the market must be given pride of place in attempting to explain why the Japanese government exhibited a strong "supply side" technological bias in its health/population quality maintenance and enhancement programs, eschewing social engineering in the entitlement field in favor of importing Western medical and public health methods and knowledge.

Not surprisingly, the technological bias was already evident during the later Tokugawa period and especially at the close of the Tokugawa era—known as the *bakumatsu* period—when Western ideas and goods starting streaming into Japan as isolationism collapsed under American pressure in the early 1950s. Dutch treatises concerning anatomy and surgical and other medical treatments had made their way into Japan through the small Dutch population residing on Dejima in Nagasaki harbor, Tokugawa Japan's sole window onto the West.[4] For instance, in

1774 Gempaku Sugita published *Kaitai Shinsho,* which describes Western concepts of anatomy. While the bakufu sponsored a special academy for the study of Chinese medicine (*kanpo*), over fifty individual fiefs set up schools to teach medicine, in some of which Dutch methods—so-called *rangaku*—were espoused (Sugaya 1976: 51 ff.). This espousal of Western methods, however, was opposed by the kanpo doctors, who managed to get the bakufu to require that all books be reviewed by the Igakuin (Academy of Medicine) before publication. Thus, while it is not untrue that the mainstream Tokugawa tradition of medicine inherited by the Meiji government was that distilled from Chinese medical theory, Western ideas had some currency during the bakumatsu period, when the bakuhan system was steadily collapsing. Indeed H. Hirota (1957) estimates that at the inception of the Meiji period around 19 percent of the doctors practiced Western medicine. In short, it was far easier for the central authorities to swing the country toward Western technical concepts of public health and medicine than toward a restructuring of the system of entitlements, which, even more than medical theory, was tied up with deeply held cultural traditions. But even in the area of medicine a strong opposition campaign was mounted by the defenders of Tokugawa traditional practice.

The new Meiji government took up the challenge laid down by the kanpo school and carried through the struggle by decisively tilting in favor of the Western school. But the issue was not quickly settled. The Meiji oligarchs proceeded on a variety of fronts, at first placing the Inuka (Medical Affairs Section) under the control of the Ministry of Education, but later renaming it the Eisikyoku (Sanitary Bureau) and placing it under the Ministry of Interior. To this agency was delegated the authority and responsibility for issuing guidelines on the standards expected of medical and public health personnel, the formulation of regulations for dealing with epidemics (with the opening up of the country to international trade after the mid-1850s epidemics became a problem), and the control over and testing of drugs. After systematic comparison between the efficacy of Western and Chinese medical practices, the Inuka bureaucrats decreed that Western concepts were to be given preference in the examination tests required of those seeking certification as doctors and in the licensing of schools offering programs in medicine (see Sugaya 1976: 45 ff.). However, since most doctors already practicing during the early Meiji period used Chinese methods, kanpo practices continued to dominate throughout the late nineteenth century, a fact that increasingly ran counter to the posture of the government.

Within the government a growing belief in the efficacy of Western medical practices—especially German practices, on which Dutch rangaku medical theory was based—led the government to not only tilt in favor of Western medical concepts in examining and training doctors but also to enthusiastically embrace the new field of bacteriology. For example, the bacteriologist Shibasaburo Kitasato, who was the first to isolate the tetanus bacillus, was sent at government expense to work with Robert Koch in Germany.

In short, the Meiji government actively exploited Western imports in the area of technological improvements—what I call supply side improvements as opposed to demand side factors such as the organization of national entitlement programs like those developed in Germany under Bismarck—to enhance population quality. In particular, through its power to impose regulations the central authorities actively pursued a policy of raising standards for medical and public health personnel. Doctors were required to register with the government, and it was decreed that examinations for certification of doctors were to follow nationally imposed guidelines (although a fully standardized national test was not introduced until after World War II). In 1899 midwives were brought under regulation and in 1915 regulations for nursing were promulgated. One of the consequences of this policy of standardizing around Western medical principles was a temporary reduction in doctors per person. The kanpo doctors aged and eventually either died in active service or retired. Due to inevitable delays in the establishment and staffing of educational and medical institutions designed to train young personnel in Western methods, replacements to the ranks of the medical profession were outpaced by the older generations of kanpo doctors. This explains why, as we see in the lefthand column of table 9, the number of doctors per capita decline between the decade 1911–1920 and the decade 1921–1930, before increasing thereafter. Even in the area of medical personnel the legacy of the Tokugawa period lingered far into the Meiji period. And because of the tradition of balkanization of entitlements central to Tokugawa government policy, gradually gathering momentum toward centralization was slowed.

But the long-run trend definitely favored centralization, and one of the most important agents of that change was the military. A case in point is the military's innovations in preventing the spread of disease in military camps due to waterborne microorganisms. During the Sino-Japanese War in the 1890s, the number of Japanese soldiers dying from infections far exceeded the number dying from war-related wounds.

Because of this the Japanese military authorities began a systematic study of why epidemics broke out in military encampments: they dispatched a research team abroad to study methods of preventing infection in foreign military organizations, finally settling on American practices, which they then methodically implemented. Henceforth they equipped all base and field hospitals with bacteriological laboratories; they made certain that every division included a sanitary detachment that carried water-testing kits; they made compulsory the boiling of water; and so forth. Thus during the Russo-Japanese War at the beginning of the twentieth century, the ratio of those dying from infection to those dying from wounds dropped to one to four. And of course, since military service was compulsory—although exemptions to military service were granted—the military innovations diffused down to the village level both through government regulation and through word of mouth.

In contrast to its assertive role in promoting the importation and dissemination of German medical and public health knowledge, the Japanese government showed remarkably little interest in adopting German innovations in the field of entitlements, for example, health and disability insurance, legislation regulating contamination in factories, and so forth.[5] The reason has already been stated: the legacy of the Tokugawa period carried with it an assumption that voluntary agreements between employer and employee lay at the center of health enhancement and that insofar as governments felt compelled to intervene, responsibility was to be exercised at the local community level. A telling illustration of this point is the protracted length of time required for passage and implementation of national mining and factory acts setting minimum safety standards, restricting the amount of overtime work, and outlawing child labor: over a half century of debate and study went into this effort. The lethargic speed at which legislation was adopted could hardly be said to be due to lack of knowledge of Western practice: during the early Meiji period the government had passed legislation assuming responsibility for the factories it directly managed. For instance, it hired French doctors in the Ikuno branch of the Government Mining Bureau to look after the health of the French technicians and miners who worked in the pits. By the 1910s mining injuries were in excess of 150,000 a year, many stemming from accidental explosions of inflammable gases, and yet stiff regulations were limited to government-managed operations. Moreover, the central bureaucracy had actually tried to draft and get Diet passage for a mining law as early as the 1880s. Finally in 1905 a mining law passed the Diet. And even more time was required to pass a factory act: even when the Kogyōhō (Factory Act) finally reached the floor of the Diet in 1910, resistance to the legislation remained widespread among the ranks of organized business who argued the "beautiful Japanese traditions" governing the relationship between employer and employee should not be subject to government intervention and regulation. And the law that was finally passed in 1911—whose twenty-five articles included banning employment of minors; stipulating that a certain number of minutes should be set aside each workday for rest; requiring that factory owners compensate employees disabled by dint of their duties in the factory; and appointing a small number of factory inspectors to investigate conditions in factories above a minimum size—was not actually implemented until 1916, almost fifty years after the Meiji Restoration.

Given the central government's bias toward supply side—technological—solutions to health enhancement and away from demand side—entitlement—approaches, the responsibility for organizing and financing public health and medical activities and for policing factories largely fell to local authorities. For this reason we should not be surprised that there is a fairly close relationship between per capita income and size and density of communities—the larger a community, the greater the economies of scale in the provision of clean water, removal of sewage, and the dispensing of medical knowledge—and per capita levels of investment in public health and medicine. In regard to this point see table 17, which gives figures for the forty-seven prefectures of Japan classified by levels of urbanization and per capita levels of medical personnel, hospitals, and hospital capacity during the late nineteenth and early twentieth century.[6] There are certainly exceptions—see the figures on maximum levels for group D—but in general the greater the level of urbanization, the greater the per capita resources devoted to enhancing health. Balkanization of health-enhancing entitlements was continuing. Now, however, the force of per capita income and the tax base and scale economies were dictating the geographic pattern, not the dual administrative bakuhan system. And that local areas, not the central authority, were carrying the main fiscal burden for these programs helps us to make sense of the very low levels of national government expenditure on social security (including public health and medicine) evident in table 16.[7] In short, many aspects of the Tokugawa heritage in entitlements were, under a new guise, being perpetuated in the Meiji period, even four decades after the bakuhan system lay in ruins.

TABLE 17

Urbanization and the Geographic Distribution of Medical and Public Health Services, 1890-1910

Group[a]	% shi[a]	Physicians 1890	Physicians 1910	Pharmacists 1910	HEHIWPC[b] 1900	HEHIWPC[b] 1910	PATCAP[c] 1890
A, avg.	27.9	99.6	86.5	13.9	12.8	13.9	253.7
A, max.	72.4	139.9	164.5	33.5	24.5	23.0	842.9
A, min.	11.2	70.4	52.7	7.1	6.1	7.2	16.2
B, avg.	7.0	90.8	64.2	4.2	16.2	19.8	105.0
B, max.	8.7	140.5	87.2	6.8	26.0	28.4	1,118.0
B, min.	5.1	41.6	38.5	1.8	4.9	6.3	7.1
C, avg.	3.8	79.9	59.4	4.8	16.5	19.2	56.0
C, max.	5.0	121.2	81.5	9.0	33.4	34.3	151.3
C, min.	2.5	52.4	40.5	1.5	2.6	5.3	16.2
D, avg.	0.0	75.1	54.5	4.0	12.2	15.9	83.9
D, max.	0.0	105.8	80.7	9.5	27.4	30.0	439.9
D, min.	0.0	15.6	28.1	0.8	0.6	2.6	7.1
Nation	9.2	85.0	67.2	7.0	14.9	17.2	103.1

(Rates per 100,000 Population)

SOURCES: Umemura et al. 1983: various tables.

NOTES: [a]Let % shi = % living in cities (shi). Then Group A (7 prefectures) has % shi greater than or equal to 10; Group B (14 prefectures) has % shi between 5 and 9; Group C (13 prefectures) has % shi greater than 0 and less than 5; and Group D (13 prefectures) has % shi equal to 0. Okinawa is included here.

[b]HEHIWPC = hospitals, epidemic hospitals, and isolation wards per 100,000 population.

[c]PATCAP = maximum capacity for patients in hospitals per 100,000 population.

ENTERPRISE AND THE FEMALE FACTORY WORKER

How bad were conditions in the textile factories? To what extent do the anthropometric measures allow us to draw conclusions about the supply pool from which the girls were drawn and about conditions in the factories themselves? Many studies, for example, E. P. Tsurumi's (1990), have documented the fact that the mills were dirty, noisy, and poorly ventilated. But how much of a toll did these conditions exact on the girls who entered the mills?

We have some evidence on both points. As a result of the passage of the Factory Act of 1911 factory inspectors began to enter the larger mills in 1916 and to file reports on an annual basis thereafter.[8] From the tables compiled from these reports it is clear enough that mortality rates among workers and former workers in the factories exceeded those for persons of comparable age in the prefectures in which the factories were located. Moreover, rates of airborne infection were especially high in the factory mills, particularly for workers who resided in factory dormitories and were thus exposed both at work and at their place of residence.

Anthropometric data for spinning mill recruits and spinning mill workers of various lengths of job tenure allow us to supplement the qualitative impressionistic accounts penned by factory workers and social critics. The data appear in table 18. On the basis of these figures we can make the following points. First, factory recruits were considerably shorter than students at all ages; second, factory recruits seem to have matured at later ages than did students (the gap in height between recruits and students decreases between ages 15 and 20); third, in comparison to students, recruits are heavier at any given height (i.e., recruits have larger body mass indexes than do students). Therefore, there is no doubt that there were rather large differences in the population quality of the supply pool from which the spinning mills drew and that of the female student population who were, in general, daughters of middle-class and well-to-do families.

But what about operatives who had worked in the mills a year or two? What was the extent of the impact of the work and living environment of the mills? As we can see from panel B of table 18, the pattern is a bit erratic, but in general the argument that more years of work in the mill adversely affected physique is not confirmed by these data. It is possible that there is a selectivity problem. That is, it is possible that the less healthy and sickly voluntarily or involuntarily left the mills early on in their work careers. But if we compare recruits with those

TABLE 18

Anthropometric Measures for Female Students, Factory Recruits, and Spinning Mill Operatives,
Ages 15-20, Circa 1910

A. Students, Fresh Factory Recruits, and Average for Spinning Mill Operatives

Age	Height (cm)			Weight (kg)			Body Mass Index		
	Student	Recruit	Operative	Student	Recruit	Operative	Student	Recruit	Operative
15	143.0	136.4	136.7	38.6	37.9	35.7	18.9	20.4	19.1
16	146.4	140.3	139.4	42.4	38.1	39.0	19.8	19.3	20.1
17	147.3	141.8	140.9	45.0	41.5	42.1	20.8	20.6	21.2
18	148.2	141.5	141.5	47.2	44.5	44.4	21.5	22.2	22.2
19	147.9	142.4	143.9	47.6	47.1	45.8	21.8	23.2	22.1
20	147.9	143.3	145.1	48.1	47.9	47.2	22.0	23.4	22.4

B. Spinning Mill Operatives with Various Degrees of Factory Work Experience[a]

Age	Height (cm)			Weight (kg)			Body Mass Index		
	0-1 yr.	1-2 yr.	2 yr.+	0-1 yr.	1-2 yr.	2 yr.+	0-1 yr.	1-2 yr.	2 yr.+
15	139.7	138.9	134.8	36.1	35.5	36.8	18.5	18.4	20.2
16	141.8	140.0	136.8	41.8	40.0	38.7	20.8	20.3	20.8
17	147.6	142.9	141.2	43.4	42.5	42.8	20.0	20.8	21.5
18	140.3	137.4	145.6	46.1	46.9	44.4	23.4	24.8	20.9
19	142.1	145.7	137.6	47.4	47.6	45.5	23.5	22.4	24.0
20	143.3	145.7	142.7	47.6	47.7	45.8	23.2	22.5	22.8

SOURCES: Kagoyama 1970: various tables.

NOTES: [a]My estimates based on figures on differential height and weight in the source. The original figures for height are in *bu* (where 10 bu = 1 shaku = 30.3 cm) and the weight figures are in *momme* where 1 momme = 3.75 gr.

having mill experience, the superiority of the latter is evident. This does not really prove the argument of the mill owners that the "beautiful Japanese relations" between employer and employee were promoting enhanced health. After all, the mills drew from some of the lowest income groups in the country and it is quite possible that the food provided to the factory girls in company cafeterias, which was designed to keep up the stamina of the workers, was superior to that eaten in poor agricultural households. But the data do not contradict the argument of the mill owners, either. Perhaps the most reasonable way to put the matter is this: by the early twentieth century socioeconomic differentials in population quality were becoming very large. This was in part due to widening income differentials as the incomes in the industrial conurbations like Tokyo and Osaka increased and those in rural areas increasingly lagged behind; the process whereby differentials widened was not rapid as economic growth was balanced, but it was steadily occurring nevertheless. But more important were the debilitating effects of physical work and disease. The female students were not accustomed to doing physical work on a regular basis, and they came from households that could afford to spend more time and resources on coping with infectious disease. In short, net nutrition was far better among the young adult student population of Japan than it was among the young adult factory worker population. And the fact that central government was relying heavily on the market and was not aggressively promoting entitlement programs to supplement market outcomes was contributing to the widening of the differential.

REGIONAL DISTRIBUTION OF YOUNG MALE HEIGHT

Without doubt, then, socioeconomic differentials were pronounced toward the close of the era of balanced economic growth. But what about geographic differentials? In light of the fact that Tokugawa period entitlements were balkanized and that socioeconomic differentials appear to be quite pronounced, would we not expect these to be substantial as well? After all, the combination of income per capita and socioeconomic differentials, coupled with the fact that some fiefs had devoted more resources to ensuring their peasants against food crises than other fiefs (which presumably left its mark on the physical growth of children at the beginning of the Meiji period), should generate regional variations in height and weight among young adults during the period of balanced growth.[9]

We can begin the investigation with regional figures on the percentage of twenty-year-old males examined in military recruitment physicals who are short (defined as 4 *shaku* or less [1 shaku = 30.3 cm]). As can be seen from map 1, the forty-six prefectures of Japan (excluding prefecture 47, Okinawa) fall into ten regions. And for these ten regions we have data for 1918 on the percentage of males examined for military recruitment examinations who are short. The percentages are given on map 2. There is considerable regional variability in shortness. It is least evident in Hokkaido—where the Ainu population, which once was genetically distinct, resides and has intermarried with the population originating from the remainder of Japan—and in the Kinki region, which includes Osaka, Kyoto, and Hyogo prefectures, which during the Tokugawa period constituted the most economically advanced area of Japan. The regions where shortness is most pronounced are the districts in the northeast of Honshu, the main island, and especially the districts just north of Tokyo. To what extent these differentials are long-standing is, of course, a matter of debate. That they may have been long-standing is suggested by the fact—for which some statistical evidence is offered in the next chapter—that the high income areas of interwar Japan tended to be those that had been heavily urbanized during the late preindustrial period.

CONCLUSIONS AND IMPLICATIONS

In this chapter I have argued that despite the wrenching changes brought on by the introduction and adaptation of Western technology and institutions in Japan over the period 1880–1920, the structure of the labor market as measured by intersectoral differentials in labor productivity and in wages did not experience sudden and dramatic upheaval. Indeed, what is most striking about the labor market is how strong is the continuity running from the late Tokugawa period into the period around World War I. However, while the composition of employment in terms of gender and wage levels relative to agriculture does not exhibit dramatic change, the size of industrial enterprises employing workers does. As a result of an increase in the number of large firms employing workers, especially in mining and in textiles, there was a gradual breakdown in the health-enhancing institutions built into labor markets during the late Tokugawa period. As these institutions crumbled there was a growing potential for a widening of the gap in population quality between various socioeconomic groups within the population of Japan.

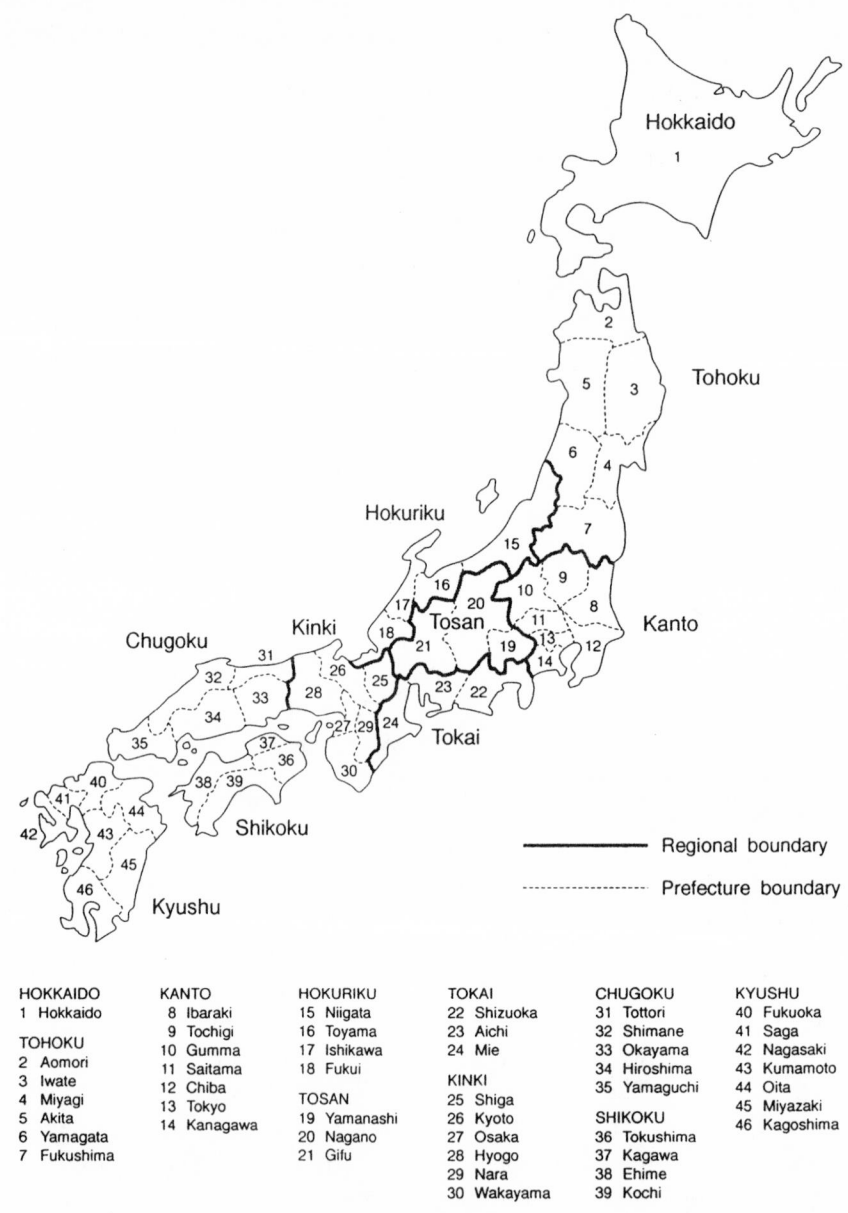

HOKKAIDO	KANTO	HOKURIKU	TOKAI	CHUGOKU	KYUSHU
1 Hokkaido	8 Ibaraki	15 Niigata	22 Shizuoka	31 Tottori	40 Fukuoka
	9 Tochigi	16 Toyama	23 Aichi	32 Shimane	41 Saga
TOHOKU	10 Gumma	17 Ishikawa	24 Mie	33 Okayama	42 Nagasaki
2 Aomori	11 Saitama	18 Fukui		34 Hiroshima	43 Kumamoto
3 Iwate	12 Chiba		KINKI	35 Yamaguchi	44 Oita
4 Miyagi	13 Tokyo	TOSAN	25 Shiga		45 Miyazaki
5 Akita	14 Kanagawa	19 Yamanashi	26 Kyoto	SHIKOKU	46 Kagoshima
6 Yamagata		20 Nagano	27 Osaka	36 Tokushima	
7 Fukushima		21 Gifu	28 Hyogo	37 Kagawa	
			29 Nara	38 Ehime	
			30 Wakayama	39 Kochi	

Map 1.   The Regions and Prefectures of Japan

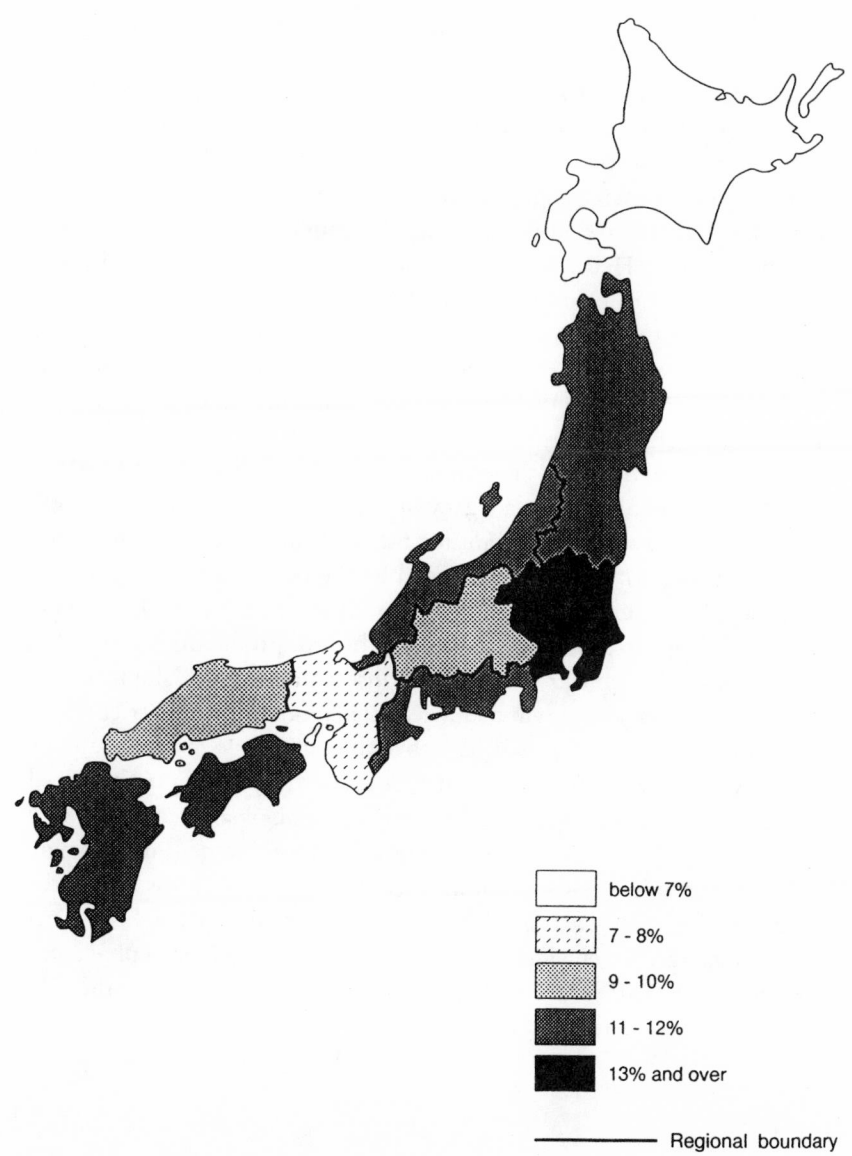

	below 7%
	7 - 8%
	9 - 10%
	11 - 12%
	13% and over

———— Regional boundary

Map 2.   Percentage of Males Examined for Military Service Who Are Short
(4 Shaku or Less) in Each Region of Japan, 1918

Moreover, the central government was not eager or able to step in and—through regulation and a program of centralized entitlements—redress the disparities in population quality that by the Meiji period were becoming increasingly apparent. The legacy of Tokugawa Japan played a decisive role in determining the course government intervention would take. Hence during the first half century of industrialization the central government concentrated its efforts on importing Western knowledge in the areas of public health and medicine and on setting standards for medical personnel, focusing its attention on supply side policies rather than on the demand side entitlement policies pioneered by Bismarckian Germany.

But the breakdown in the population quality-enhancing institutions of the market coupled with a legacy of balkanized entitlements that was becoming increasingly irrelevant for the new challenges of industrializing Japan was not to go unchallenged by the poorer and less economically favored groups in Japan. Moreover, after 1920 balanced economic growth disappeared. A surge in heavy industrial production occurred in the wake of World War I in part because embargoes of shipments to Japan among the belligerent countries acted as a nontariff barrier to imports of manufactured goods, thereby stimulating the rapid expansion of heavy industry. With the emergence of dualistic or unbalanced growth the potential for an even greater widening of socioeconomic and geographic differentials in population quality loomed increasingly large. The result was rebellion in the countryside and pressure for the creation of a new system of entitlement insurance to replace the defunct Tokugawa entitlement institutions. The story of emergent unbalanced growth and the attendant political struggle over entitlements is the subject of the next chapter.

Enterprise, Community, and Human Growth in an Era of Unbalanced Economic Growth, 1920–1940

A CHANGING MARKET ENVIRONMENT

After World War I, heavy industry, which hitherto had constituted a minuscule sector submerged by a sea of agricultural and light industrial enterprises, emerged as the driving force, the engine, of Japanese economic growth. As a result the era of balanced growth ended. And a new epoch characterized by wide and widening differentials in wages and labor productivity between the heavy industrial sector and the rest of the economy commenced.

The new market environment brought down once and for all the curtain on the system of Tokugawa entitlements. Why? Recall that Tokugawa entitlements had supplemented a set of market institutions that had automatically built in—through incentives—a considerable amount of health protection for employees. Indeed, in the case of the majority of workers in Tokugawa Japan the "employer" and the "employee" were really inseparable since the enterprise for which the latter worked was the family into which he or she was born. The entitlement system was therefore oriented to providing insurance for households that were also enterprises so that they could continue to stay in operation. Since most enterprises engaged in rice cultivation and rice was ultimately the basis for taxes and for stipends for the samurai, entitlements were focused around rice. Entitlements basically involved transfers of rice and were aimed at dealing with short-run fluctuations in supply and demand at the local level, with market price serving as a barometer for this supply/demand balance or lack of balance. Most ikki protests, most conflicts between landlord and tenant, ultimately had their origins in local conflicts over the distribution of rice. Entitlements were balkanized and therefore conflicts over them were also local.

But the trends underlying and leading up to the new market environment made a mockery of the Tokugawa system of entitlements. First of all, with the growing breakdown within the labor market of the industrial sector of enhancement of population quality, the assumption that matters could be left to enterprises was no longer credible. And even more important was the decline in the proportion of the labor force working in agriculture, and within the farming sector itself the decline in farming income relative to total income. As long as domestic rice production expanded rapidly enough to satisfy a population that, increasingly, was engaged in pursuits other than producing rice, the government allowed the farmer to enjoy the benefits of a healthy expansion in demand in the form of favorable—favorable to farmers, that is—movements in terms of trade of agricultural products vis-à-vis manufactured products. But as best-practice Tokugawa agricultural techniques diffused throughout the countryside, growth in agricultural productivity began to falter. It became apparent to a government concentrating its focus on the building up of production of iron and steel, chemicals, and transportation equipment that relying on domestic sources for foodstuffs might hamper the speed with which the nation could turn toward nonagricultural pursuits. The Rice Riots of 1918 heralded the end of an agricultural policy aimed at solving the problem of food supply for urban industrial regions through technological progress in domestic agriculture. The government began to aggressively promote the importation of foodstuffs, especially rice, from its colonies, Korea and Taiwan. In short, the central government adopted a policy that inevitably depressed either the level or growth of domestic rice prices relative to manufacturing output price levels and growth rates.

The political tilt toward industrial districts created havoc with the traditional concept of entitlements for two very important reasons. It made clear that the government was now sufficiently satisfied with domestic conditions in farming—of the ability of agricultural villages to avert subsistence food crises—that it could turn the principal focus of its food policy away from promoting domestic productivity to guaranteeing to the growing legion of industrial workers that prices for basic food staples would remain fairly stable relative to nominal wages. In effect, the government was trying to solve the growing population quality problem caused by the gradual breakdown in health-enhancing labor market institutions within industry with its rice import program. In weighing the needs of the industrial sector against those of the agricultural sector the central authorities came down on the side of industry, effectively transforming the rice price question from an agricultural into an industrial entitlement issue. Moreover, partial justification for this policy twist was the fact that agricultural crises occurred with diminishing frequency and ferocity. And a corollary of this was the fact that tenants benefited less and less from the entitlement insurance traditionally offered by landlords since the probability of harvest failure was steadily in decline. Thus the two mainstays of Tokugawa entitlement insurance, which had been falling into disuse even during the last decades of the balanced growth era, now ceased to exist altogether. The result was rural rebellion—limited perhaps, but rebellion nevertheless—as the rural sector, especially its poorer segments, began to aggressively voice its discontent with the attention shown to the industrial sector and the almost exclusive attention now being devoted to building up a new system of entitlements to cope with the social problems of manufacturing. The increasing policy of according wage premiums to male workers in heavy industry served to further fan the flames of rural discontent.

The wrenching changes experienced in the structure of labor markets can be seen in table 19. First consider wage differentials in panel C of that table. The premium enjoyed by male industrial workers both over male workers in agriculture and over female industrial employees sharply and decisively increases during the interwar period, especially during the 1930s. In response to the changes in relative wage ratios for males and females in industry and in agriculture, which in turn reflects the surge in demand for males in heavy industry, farm families began to change the sexual composition of their labor forces. This is apparent in panel A of the table: the number of male workers declines with some fluctuation until the mid-1930s and then precipitously drops thereafter as the number of female workers increases, most dramatically after the mid-1930s. Female workers, who constituted the backbone of the low-wage/low-skill light industrial labor force—and increasingly this was the case as a glance at the percentages of cotton spinning hours worked by females confirms—were in a sector of manufacturing that was declining in importance relative to high-wage/high-skill heavy industry, which was becoming the new growth engine. The expansion of heavy industry as represented by machinery production and the contraction in textile production is clearly seen in the bottom right-hand portion of

TABLE 19

Selected Economic, Social, and Demographic Characteristics of Interwar Japan[a]

A. Agriculture

	Workers per House in Farm Households				Properties of Arable Land (1934-1936 = 100)[b]				
Period	No. (000s)	Male	Female	Female/male	Area	Price	Rent (Re)	Productivity (Pr)	Pr/Re
1921-1925	5,534	1.38	1.14	.82	97.6	139.7	135.6	110.6	104.3
1926-1930	5,581	1.37	1.15	.84	97.2	131.9	106.3	84.7	100.7
1931-1935	5,602	1.40	1.16	.83	99.2	96.8	84.6	156.4	117.8
1936-1940	5,535	1.29	1.26	.98	100.4	130.6	131.2	124.5	109.6

	Labor in Workdays (WD) and Real Productivity per Workday (LP)[c]			Wages or Fertilizer Prices Relative to Land Rent[d]			Fertilizer or Machinery Input per WD[e]		Terms of Trade[f]
Period	WD	Male WD	LP	Wages	NITF	PHOF	Fertilizer	Machinery	Agric./Manuf.
1921-1925	159	183	171	132.2	140.5	99.3	10.1	48.9	1.01
1926-1930	151	173	187	152.9	130.5	115.4	12.6	55.1	.78
1931-1935	160	183	190	114.2	106.1	112.7	13.3	57.7	1.03
1936-1940	163	192	191	121.0	88.4	132.7	15.7	60.6	1.14

NOTES: [a]Some series (in all three panels) are for 1880-1889, 1890-1899, etc.

[b]Indexes for rents, prices, and productivity of land are based on nominal figures and pr/re indicates the relative level of nominal land productivity (value added net of depreciation on capital assets) relative to an index of nominal land rent.

[c]"Male WD" refers to male equivalent workdays. Index for LP is with 1880 = 100. Here 1921-1925 figure is for 1925, 1926-1930 figure is for 1930, 1931-1935 figure is for 1935, and 1936-1940 figure is for 1940.

[d]Ratios of indexes (all indexes having 1934-1936 = 100) with land rent index as the denominator. NITF = nitrogen and PHOF = phosphate fertilizer.

[e]Fertilizer and machinery inputs in 1934-1936 prices and per 100 workdays.

[f]Price index of agricultural goods divided by price index for manufactured goods (1934-1936 = 100).

TABLE 19 *continued*

B. Income and Consumption per Capita and Landlord-Tenant Disputes

Period	Landlord-Tenant Disputes[g]			Income and Consumption per Capita[h]					
	Average no.	% Rent Reduction	% Tenancy Continuation	GDPPC	CONPC	FOODPC	MEDPC	EDRPC	Government Expenditure[i]
1921-1925	1,885	96.4	3.6	208.4	174.1	100.5	6.4	12.7	4.4
1926-1930	2,316	73.9	26.1	214.8	178.7	98.8	6.9	13.7	3.4
1931-1935	4,697	53.8	46.2	225.7	181.9	95.5	10.2	13.4	2.7
1936-1940	4,607	51.0	49.0	286.2	193.6	93.7	11.1	15.2	3.5

C. Relative Sectoral Wage Levels, Structure of Production, and Female Labor Input in Manufacturing

Period	Female/Male Wage Ratio		Agricultural/Manufacturing Wage Ratio		Urban Population[j]	PPI[k]	Hours, female[l]	% of Manufacturing Output	
	Agriculture	Manufacturing	Males	Females				Textiles	Machinery
1921-1925	.76	.45	.75	1.26	11.1	52.0	75.8	45.3	12.2
1926-1930	.80	.41	.65	1.27	11.8	50.3	77.7	45.6	10.8
1931-1935	.75	.32	.45	1.03	18.3	47.5	82.9	39.7	12.1
1936-1940	.79	.33	.53	1.30	20.0	45.3	87.9	30.0	22.2

SOURCES: Various tables from Hayami 1975; Japan Statistical Association 1987; Ohkawa and Shinohara 1979; Umemura et al. 1966; Umemura et al. 1988; Waswo 1977.

NOTES:

gPercentages of rent reduction and of continuation of tenancy or compensation.

hAll figures in 1934-1936 prices. GDPPC = gross domestic product per capita; CONPC = total consumption per capita; FOODPC = food consumption per capita; MEDPC = expenditure on medicine and personal care per capita; EDRPC = expenditure on education and recreation per capita.

iPercentage of central government expenditure for social security, including public health and medicine.

jPercentage of the population living in the six big cities of Tokyo, Yokohama, Nagoya, Kyoto, Osaka, and Kobe. Figures for 1921-1925 are for 1925, figures for 1926-1930 are for 1930, etc.

kPPI = percentage of the gainfully employed population (both sexes) in primary industry (agriculture and forestry).

lPercentage of hours supplied in cotton spinning by female workers.

panel C. In short, a new labor market structure was emerging and an old labor market that had much in common with that of late feudal Japan was dying.

But drastic structural change was hardly limited to labor markets. It was also apparent in product and in capital and land markets. Consider the sharp fluctuations in the terms of trade between agriculture and manufacturing: during the late 1920s the rice price and the price level of a general bundle of foodstuffs sinks; and then it soars during the late 1930s. The uncertainty of getting an acceptable harvest had now been replaced by the uncertainty of the income one could secure from that harvest. Among the industrial products declining in price relative to agricultural prices is the output of the chemical fertilizer industry. Note from panel A of table 19 that the prices of nitrogen-based fertilizer fall relative to land prices while wages rise—with some fluctuation—relative to land. A rise in the cost to farming families of labor relative to land brought on by a vigorous expansion in demand for labor within industrial production should encourage households to substitute factors for labor that are becoming, in relative terms, comparatively cheap. And trends of this sort are evident from panel A: fertilizer and machinery inputs are increased (the changing terms of trade encourage this after the early 1930s) and labor input in terms of number of workers is decreased. Because of a secular trend toward improvement in population quality, adult workers are better able to work and hence workdays per worker increases somewhat, but labor productivity increases even faster, each hour of work garnering more real output because of the quality improvements in the labor input and because labor now works with more fertilizer and with more machinery. In short, interwar agriculture was by no means incapable of productivity growth.

But while it was not incapable of generating growth, the returns to investments in agriculture were now, in the new market environment, far less potentially remunerative than those that one might secure from industrial investments. The rapid expansion of the six big cities evident from panel D is testimony to the feverish activity centered around new heavy industrial enterprises in the major metropolitan conurbations. This is one reason wealthy landlord families—the very households that pioneered the diffusion of best-practice techniques and the construction of social overhead capital in the countryside like roads and drainage ditches—now turned away from rural pursuits and began to invest funds and time and energy in the newly developing industrial sector. Hence in the new market environment landlords had little, if anything, to offer since they were no longer spearheading the drive for improvements in the countryside; they were not even particularly interested in the problem of how to redress the growing imbalance in entitlements as they saw their fate now tied increasingly to investments in urban industrial enterprises, and moreover their entitlement insurance was hardly needed. Is it surprising, therefore, that their tenants now turned against them?

CONTINUITY AND CHANGE IN REGIONAL DIFFERENTIALS IN POPULATION QUALITY

Looming over the striking restructuring of Japan's labor market was a growing military crisis; indeed, the military crisis overshadowed and conditioned the economic transformation and played an important role in shaping the proposals for a centralized entitlement program to replace the defunct Tokugawa scheme. The deepening military crisis that saw Japan expand its de facto occupation of the mainland of Asia from its colonial base in Korea to control over Manchuria and portions of China and that brought it step by step into conflict with its great Pacific rival, the United States, conditioned the economic transformation because it sped up the demand for heavy industrial output like chemicals and transportation equipment, for instance, battleships and armed vehicles and airplanes. And it conditioned the economic transformation because it increasingly drew the government into a imperialist strategy that presumed that the colonized areas of Asia would serve as suppliers of foodstuffs like rice and raw materials for the industrial heartland in Japan. And it conditioned the demand for a centralized program of entitlements because the growth in military activity increased the demand for healthy soldiers; that is, it increased the demand for population quality. Hence those sectors of the rural economy concerned about the growing market-driven disparities between urban income and opportunities and rural income and opportunities found a listening ear among the militarists who increasingly directed national affairs as Japan entered the 1930s. And insofar as it was perceived that population quality depended on household investments, which in turn depended on the market opportunities and the entitlements available to households, the problem of redesigning entitlements appeared all the more critical to decision makers in Tokyo.

With the military needs of the central government in mind, let us consider geographic differentials in population quality as measured in terms of height and weight on military recruitment examinations. After all, these were the figures that the armed forces assiduously collected in order to see which districts were supplying the best potential soldiers and which districts the worst. Had the military planners in the mid-1930s drawn a prefectural map of Japan and shaded in more darkly those districts with the higher proportions of short—stunted—recruits, they would have found the pattern displayed in map 3. As can be seen from a comparison of this map with map 2, many features of the 1918 pattern continued to persist in the interwar period. For instance, the regions with the smallest proportions of short examinees continued to be Hokkaido and the Kinki district, including Osaka, Hyogo, and Kyoto and the great cities of Osaka, Kobe, and Kyoto. But the availability of prefectural data for 1934 allows us to pick out some patterns that either did not exist in 1918 or are not evident in 1918 because of aggregation. For example, Tokyo prefecture, which consists of the ward areas of the city of Tokyo proper and its immediate semiurbanized and semirural suburbs, stands out as producing few short examinees within a northeastern region that, relative to southwestern Japan, tended to have a large proportion of short individuals as it had in 1918. Nagasaki prefecture on the southern island of Kyushu, which includes the city of Nagasaki, off of which is Dejima, the island occupied by the Dutch during the Tokugawa period, is also a supplier of mainly nonshort examinees. Now the map is primarily of interest because it gives us useful information about the nature of, and particularly the perpetuation of earlier, geographic differentials in population quality. For instance, it makes clear the superiority of the major urbanized industrial zones relative to the hinterland, especially the northeastern hinterland. But this map is also of interest because it shows us the nature of the social problem confronting Japan's military leaders: How were the areas lagging in population quality to be brought up to the standards set by the high-income industrial districts?

And it should have been clear to the military leaders that relying on the market alone would not help close the gap. Consider table 20. This table gives means and maxima and minima for groups of prefectures classified by proportion of the total labor force in primary industry averaged for the years 1920, 1930, and 1940 (PPI). Group A is the most agricultural and group D is the least agricultural. As can be seen, not only are the per capita income levels far higher in group D than in group A, but absolute increases in per capita income are greater in groups C and D than they are in groups A and B. Moreover, there is a large

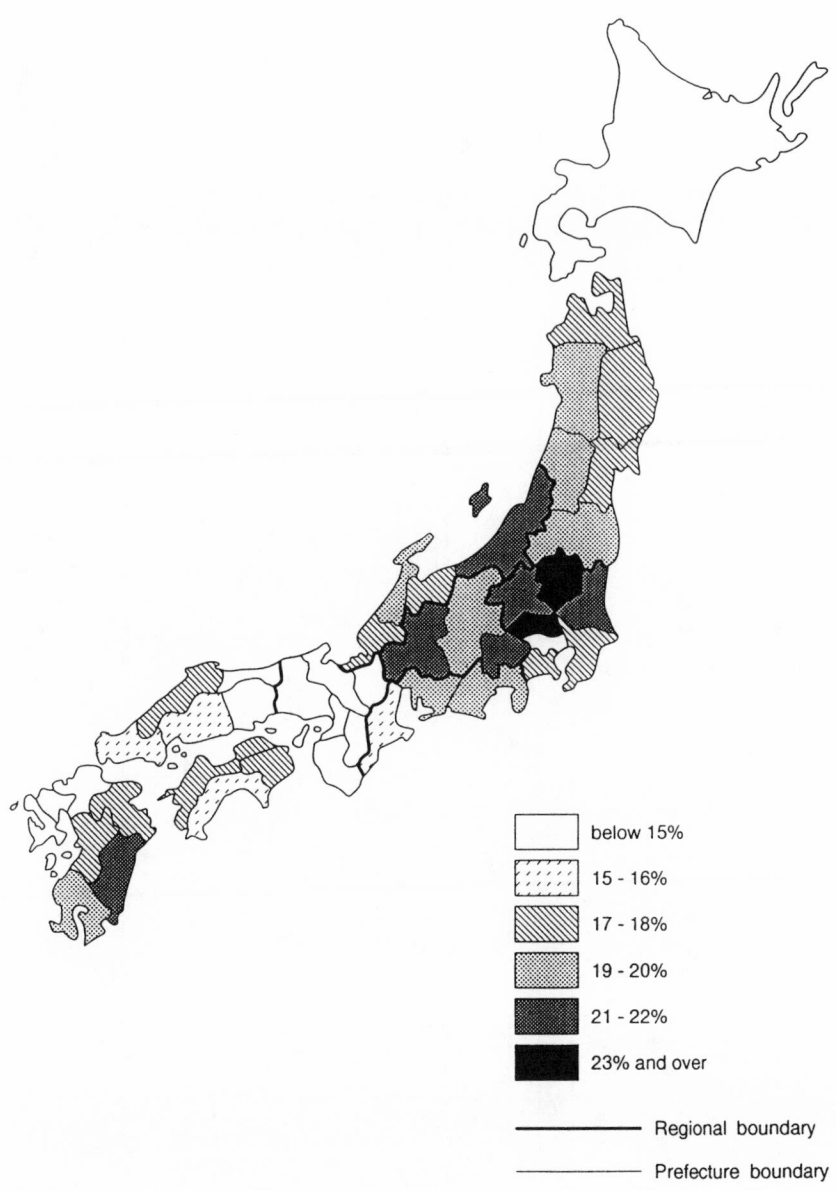

	below 15%
	15 - 16%
	17 - 18%
	19 - 20%
	21 - 22%
	23% and over

——————— Regional boundary

——————— Prefecture boundary

Map 3. Percentage of Males Examined for Military Service Who Are Short
(150 cm or Less) in Each Prefecture of Japan, 1934

TABLE 20

Income, Public Health and Medicine, and Anthropometric Measures for
Military Recruits in the Prefectures of Interwar Japan

A. Relative Income, an Index for Public Health and Medicine, and Percentage Tall and Short

| PPI Group[d] | RYPC and Change[a] | | PHMEDI, 1927 and 1935 Weights[b] | | | Percentage Tall or Short[c] | |
| | | | Based on 1927 Weights | | 1935 Weights | | |
	1930	1920/1930	1927	1935	1935	% Tall	% Short
A: mean	78.6	84.1	62.0	79.8	66.3	4.2	18.5
A: max.	93.5	286.1	88.5	108.6	81.3	6.5	21.8
A: min.	59.0	-112.3	41.2	60.4	51.7	3.3	12.1
B: mean	84.5	64.7	74.5	98.5	78.5	4.2	18.8
B: max.	107.5	488.5	104.7	130.6	98.3	7.5	25.2
B: min.	62.8	-294.7	51.6	61.3	53.6	2.7	13.2
C: mean	104.7	188.2	87.0	106.8	86.1	5.3	15.9
C: max.	121.9	427.7	98.5	128.5	102.5	6.7	19.8
C: min.	77.8	-155.3	65.5	91.9	73.3	4.0	13.5
D: mean	168.8	110.2	145.8	192.7	136.7	6.0	14.9
D: max.	238.7	596.3	235.2	297.3	211.5	7.6	19.9
D: min.	113.9	-336.8	97.5	134.7	106.0	3.9	11.2

TABLE 20 continued

B. Average Heights, Weights, and Body Mass Index

PPI Group[d]	Average Height and Gain (cm)		Average Weight and Gain (kg)		BMI and Gain	
	1930	1926/1937	1930	1926/1937	1930	1926/1937
A: mean	159.7	+0.6	53.4	+0.6	20.9	+0.1
A: max.	161.2	+0.8	54.9	+1.1	21.5	+0.3
A: min.	159.2	+0.3	52.4	+0.3	20.7	-0.1
B: mean	159.6	+0.9	52.6	+0.6	20.7	+0.2
B: max.	161.5	+2.4	54.5	+2.1	21.3	+0.5
B: min.	158.3	-0.1	51.2	+0.1	20.3	-0.3
C: mean	160.1	+1.0	53.1	+0.6	20.7	-0.3
C: max.	160.7	+1.3	55.2	+1.2	21.6	+0.4
C: min.	159.6	+0.3	52.0	+0.2	20.3	-0.2
D: mean	160.4	+1.2	52.2	+0.9	20.3	+0.6
D: max.	161.3	+1.5	52.8	+2.2	20.6	+0.5
D: min.	159.7	+0.9	51.3	+0.2	20.0	-0.3

SOURCES: Japan Statistical Association 1988: table 21-20 (pp. 182-193); Nihon Naikaku Tōkeikyoku (various years): various tables.

NOTES:

aIncome per capita relative to the unweighted mean for the prefectures which is set at 100.

bThe index based on 1927 weights is calculated by equally weighting indexes (with the national level value set at 100) for hospital beds per capita, doctors per capita, dentists per capita, pharmacists per capita, and nurses per capita. For the 1935 index two additional indexes—dental clinics per capita and general clinics per capita—are added (all seven indexes receiving equal weight).

c% tall is % 170 cm or over; % small is % 150 cm or less.

dPPI = percent labor force (both sexes and averaged for 1920, 1930 and 1940) in primary industry. Group A (10 prefectures) has PPI greater than or equal to 65%; group B (20 prefectures) has PPI between 55% and 64.9%; group C (9 prefectures) has PPI between 45% and 54.9%; and group D (2 prefectures) has PPI less than or equal to 44.9%. "max." = maximum; "min." = minimum.

disparity in the per capita index of public health and medicine (PHMEDI) for 1927—with group D prefectures far more favored than those in groups A and B—and the disparity also seems to grow larger over time (between 1927 and 1935). Again dependence on primary industry is positively associated with high levels of child/youth labor services. So disparities in all three of the factors that in chapter 2 we found were important in shaping trends in population quality—income (hence gross nutritional intake), public health and medicine, and child/youth labor services—are correlated with each other in the cross section. And as has already been discussed in the context of map 3, and as can be further seen in the right-hand columns of panel A and panel B of table 20, differentials in population quality mimic these differentials in net nutritional intake. Moreover, gains in height, weight, and BMI appear—not dramatically, but slightly—to be greater in the districts with higher levels of population quality. In short, the Japanese military authorities were faced with a serious problem of how to eliminate differentials in population quality between advanced industrial districts and the rural hinterland, especially in the Northeast. And the market was not eliminating these differentials. Indeed, because of the strong economies due to transportation costs, heavy industry was concentrating more and more in the great urban conurbations around Tokyo-Yokohama and Osaka-Kobe-Kyoto; hence if nothing was done in the policy arena, disparities in height and weight for military recruits were likely to increase, not decrease.

For these reasons Tokyo was becoming concerned about the lack of an adequate entitlement program in the countryside. And for this reason alone it is not surprising that the ministries in Tokyo began to consider or experiment with price supports for rice, a two-tier system of governmental purchase of rice under which marginal tenant farmers would receive higher prices than other farmers, a wholesale land reform that would do away with absentee landlordism, and some form of insurance for farm households administered through local rural cooperatives. But the interest of the central government in moving in the direction of erecting a new set of centrally administered and organized entitlements to replace the balkanized system inherited from the Tokugawa period is not simply due to long-run military concerns. It is also attributable to something far more immediate and pressing: the danger of wholesale rebellion in the countryside. With greater salience than any other single sequence of events, the outbreak of rural rebellion during the interwar period demonstrates the great importance of the voicing of demand for entitlements in conditioning the historical development of population quality within twentieth-century Japan.

TENANCY, PADDY PRODUCTION, POPULATION QUALITY, AND AGRICULTURAL LABOR PRODUCTIVITY

Rural revolt in interwar Japan centered around confrontations, at times violent, between tenants and landlords. The institution of tenancy was already well established by the latter part of the Tokugawa period, especially in the commercially advanced Kinki district. However, it was not until 1871 that the government began to issue deeds guaranteeing ownership to the individuals assuming the responsibility for paying taxes on the land. And it was not until 1872 that the formal Tokugawa ban on land sales was actually abolished. In 1873 the Land Tax Revision was promulgated, converting taxes from payments in rice to payments stipulated in monetary terms and based on the estimated value of land. Despite the conversion of taxes to monetized payments, however, under the typical arrangement between landlords and tenants rents continued to be paid in terms of a stipulated volume of rice or in terms of a proportion of the rice crop on the land worked under tenancy.

How prevalent was tenancy? Y. Nishida (1986) estimates that in 1874 30.6 percent of cultivable land was in tenancy and that the percentage rose thereafter, reaching 40 percent in 1887 and 45.4 percent in 1912. A similar story is told by the following figures on the distribution of farm households according to landownership status for 1883–1884 and 1912 (PO/PT stands for part-owner/part-tenant):

	1883–1884	1912
Owners	37.4%	32.5%
PO/PT	42.9%	39.8%
Tenants	19.7%	27.7%

In short, over the course of the balanced growth era tenancy grew. Perhaps this growth was due to the fact that the land tax was now collected in a monetized form, and therefore during the Matsukata Deflation of the 1880s when rice prices dropped, many marginal farm households were forced to sell their land. Or perhaps it was due to the fact that reclaiming land for cultivation was an attractive investment opportunity during the era of balanced growth and therefore enterprising merchants and risk-taking farmers were encouraged to invest in the development of land that they could then rent out to tenants. In any case, because so much cultivated land was farmed under tenancy, by the time Japan entered into the era of unbalanced growth the potential for a major political struggle between landlords and tenants was well developed.

Beginning in the early Meiji period tenants here and there had taken collective action potentially directed at landlords: they formed tenant unions.[1] The first tenant union was established in 1875, and the number gradually grew after that. However, after over forty years of organizing, in 1917, the total number of unions did not even reach two hundred. But then, all of sudden, the number of unions and their activity proliferated. As can be seen in panel B of table 19, landlord-tenant disputes increased at a fever pitch during the 1920s. Over the course of the interwar period both the quantity and the character of the disputes changed. During the early 1920s tenants appear to have been the main initiators of disputes as the vast number centered around demands for rent reductions. However, during the later 1920s and the 1930s landlords took the offensive in many areas and the proportion of disputes involving continuation of tenancy or compensation due to termination of a landlord-tenant contract soared. Hence it is clear that landlord-tenant conflict became especially pronounced precisely at the time when the balkanized Tokugawa entitlement system was collapsing and a centralized replacement emanating from Tokyo had not yet filled the vacuum. It is my view that the outbreak of rebellion in the countryside was not merely due to conflict between tenant and landlord but rather was the voicing of a demand on the part of villagers for a new system of entitlements to replace the defunct Tokugawa system. Because landlords had once been a source for traditional entitlement insurance, they often bore the brunt of the attack on the old system. But the rebellion in the countryside was only in part a rebellion designed to initiate land reform. It was only part of a much broader demand, namely, demand for a system of entitlements consistent with the realities of a new market environment.

The rural village, especially the poorer stratum of the village represented by the marginal owner and the smaller tenant, was being left behind: the new market was leaving it behind; the government was leaving it behind. That was the fear that fanned rural discontent. With the abolition of feudal classes, upward social mobility became a reality for many poorer but ambitious farmers. In the balanced growth era of the

Meiji period agriculture offered opportunities comparable to those of-fered by industry; hence the low-income hardworking tenant house-hold, by dint of hard work and a detailed knowledge of farming tech-niques or by dint of the assistance of a landlord family interested in increasing its own rental income in tandem with its tenant's income, could advance economically and socially. But in the new market envi-ronment favoring industry—especially heavy industry—the small farmer and especially the small tenant farmer watched with growing anxiety as the relative agricultural wages and returns on investment fell.

The key point to keep in mind in understanding why rural rebellion exploded during the interwar period is that it was triggered by a sense that the government was abandoning the rural sector to the whims of an increasingly hostile and unpredictable market in which less prosper-ous farmers expected to experience relative losses of income and status. The decision to commence wholesale imports of rice from the colonies was a symbol of the apparent lack of interest by the central authorities in building up a new system of entitlements to replace those that in the late Tokugawa period had stabilized local rice markets in times of dearth. It is important to stress that *relative* income and status were at the heart of the matter, not absolute levels of income. For instance, con-sider the figures in table 21 on household membership and sex compo-sition, relative income per household worker and relative agricultural income per household worker, and growth rates in real income per household worker and growth rates in real expenditure per household member. As can be seen from the figures that compare levels and growth rates for large and small owner households, large and small part-owner/part-tenant households, and large and small tenant house-holds, income was growing in real terms throughout the 1930s for all six groups.[2] But the gaps between the groups were also slightly growing in most, but not all, of the measures looked at here. In short, even with the terms of trade moving in a favorable direction, rural households, es-pecially marginal tenant households, had a growing sense of insecurity: the tenant household did not enjoy security of land ownership and to this insecurity was added the growing insecurity of the rice market, which was now being affected by conditions of production in distant colonies like Korea and Taiwan.

Central to the concerns over a deterioration in relative incomes was a concern about the long-run—by long run I mean a period of several gen-erations—economic and social status of the ie. For "feedback" of income into expenditures on education, food, and direct health maintenance,

TABLE 21

Tenancy, Scale of Operation, Investments in Labor Efficiency Improvements, and Productivity Growth in Rural Japan, 1931-1940

A. Household Composition and Income per Hours Worked, Levels, and Growth Rates

	Agricultural Households by Tenancy Status and Scale of Operation					
	Owners		Part-Owners/Part-Tenants		Tenants	
	Large Farm	Small Farm	Large Farm	Small Farm	Large Farm	Small Farm
HM, 1931-1935[a]	6.7	5.4	6.6	5.0	6.6	5.4
HM, 1936-1940[a]	6.8	5.6	6.7	5.3	6.7	5.7
SRHM, 1931-1935[a]	95.8	119.5	97.7	91.6	97.5	118.3
SRHM, 1936-1940[a]	111.9	118.9	103.9	105.8	97.2	122.7
RAGYPW, 1931-1935[a]	100.0	92.2	89.9	78.1	69.8	60.8
RAGYPW, 1936-1940[a]	100.0	84.9	87.9	76.3	72.2	54.5
RYPW, 1931-1935[a]	100.0	98.9	92.3	83.8	74.7	71.1
RYPW, 1936-1940[a]	100.0	92.3	91.0	85.8	76.8	67.1
GREXPHM, 1931-1940[a]	+2.9	n.e.	+3.5	+3.0	+3.5	+3.6
GRYPHW, 1931-1940[a]	+8.0	+5.9	+7.5	+7.6	+7.7	+6.9
GRAGYPHW, 1931-1940[a]	+8.7	+7.7	+9.2	+7.7	+9.7	+10.3
GRPHWH, 1931-1940[a]	-3.5	-14.9	-3.7	n.e.	-7.2	n.e.

TABLE 21 continued

B. Income Elasticities for Expenditures Improving Future Labor Productivity

| | Agricultural Households by Tenancy Status and Scale of Operation | | | | | |
| | Owners | | Part-Owners/Part-Tenants | | Tenants | |
	Large Farm	Small Farm	Large Farm	Small Farm	Large Farm	Small Farm
	Elasticities of FDE, EHE, and FDEEHE[c]					
FDE[b]	+.33*	+.70*	+.58*	+.36*	+.43*	+.64*
EHE[b]	+.88*	+.50	+.46	+.81*	+.38***	+.87*
FDEEHE[b]	+.41*	+.66*	+.56*	+.42*	+.43*	+.60*
	Impact on Labor Productivity of Previous Year FDEEHE and FERPHW Expenditures[d]					
Correlation[b]	+.97	+.64	+.92	+.95	+.84	+.86
FERPHW[b]	+1.06*	+.96**	+1.07*	+.82**	+.88*	-.37
FDEEHE[b]	+2.28*	+.99**	+1.63*	+1.85***	+1.98*	-.14

SOURCE: Nihon Nōrinshō Nōmukyoku 1953: various tables.

NOTES: [a]HM = resident household members per household; SRHM = sex ratio of resident household members (females per 100 males); RAGYPHW = relative agricultural income per household agricultural worker with figure for Owner, Large Farm type = 100; RYPHW = relative income per household member, with figure for the Owner, Large Farm type = 100; GREXPHM = annual growth rate for real expenditures per resident household member; GRYPHW = growth rate for income per hour worked; GRAGYPHW = growth rate of agricultural income per agricultural hour worked; GRPHWH = growth rate of percentage of hours worked which is supplied by hired workers and not resident household members. Growth rates estimated from regressions with the logarithm of the variable regressed against time and a constant.

[b]FDE = per household member expenditure on food and drink; EHE = per household member expenditure on education and health; FDEEHE = sum of FDE and EHE; FERPHW = value of fertilizer input per hour worked; "Correlation" refers to the correlation between the two independent variables FDEEHE and FERPHW.

[c]Estimated from equation of form

$$\log(E_i) = a_0 + a_1 \log(RYPW) + a_2 \log(SRHM) + \varepsilon$$

where E_i is expenditure on category of goods i (e.g., E_i = FDE, etc.).

[d]Estimated from equation of form

$$\log(RYPW_t) = b_0 + b_1 \log(FDEEHE_{t-1}) + b_2 \log(FERPHW_{t-1}) + \varepsilon.$$

*Significant at the 1% level (two-tailed test).

which, by improving population quality, had the potential to enhance future income-generating potential for the members of the family, was important. This is the message of panel B of table 21, where I report on time series regressions for the 1930s of two forms. The first set of regressions gives for each type of farm households estimates from log-log regressions for the elasticities of per-household member expenditures on population quality-enhancing items—on food and drink, on education and health, and on the total of the two—with respect to real income per household worker (where the symbol RYPW here refers to real income adjusted for price changes, not to relative income as it does in panel A of table 21). As can be seen, the elasticities appear to be high. Moreover, as can be seen from the bottom part of panel B where I report on a second set of regressions for each type of farm household in which, controlling for the sex ratio, labor productivity in a given year is regressed against per household member expenditures on food, education, and health and per work hour real expenditure on fertilizer in the previous year, the more a household put into building up its population quality and its fertilizer input, the better off it was likely to be in future years. In short, because of the feedback effect of investments in population quality on future labor productivity, any increase in income disparities was feared by rural households. And it must be remembered that marginal farmers were concerned about two types of income disparities: the gap between urban and rural households, which was growing; and the gap between large owner households and marginal producers, which was not shrinking. And since large owner households enjoyed incomes sufficient to allow them to educate their children in higher echelons of the educational system that opened the doors to employment opportunities in heavy industry, they could escape the "squeeze" arising from the new market environment. But in general tenant households could not expect to escape this squeeze and therefore they sought redress in the form of new government entitlement programs.

A key part of this argument rests on the notion that in relative terms tenant households were suffering a deterioration in population quality. That is to say, even though they were enjoying improvements due to the strong secular trend, their relative position was deteriorating. In regard to this point consider the findings reported in table 22. In panel A I report on cross-sectional regressions using logarithms of all variables for 1927, 1930, and 1935 (the observations used in both panel A and panel B regressions are for the thirty prefectures of Japan whose average level of proportion of labor force in primary industry for 1920, 1930, and

TABLE 22

Paddy Production, Tenancy, and Anthropometric Measures for Military Recruits from the Agricultural Prefectures[a]

A. Levels (All Variables, Independent and Dependent, in Logarithmic Form)

Dependent Variable	Constant	PHMED[b]	Paddy[c]	Tenancy[d]	Income[e]	Adjusted R^2
Percentage tall	-1.15 (-.40)	+.44** (2.37)	+.51* (3.17)	-.44*** (-1.69)	+.03 (.10)	.41
Percentage small	+3.09 (1.58)	-.37* (-2.96)	-.40* (-3.61)	+.41** (2.35)	+.13 (.78)	.46
Average height, 1927	+5.08* (90.95)	+.01*** (2.95)	+.01* (3.85)	-.01 (-2.10)	-.004 (-.735)	.38
Average height, 1935	+5.06* (104.44)	+.01* (2.95)	+.01* (3.85)	-.01** (-2.54)	-.003 (-.591)	.48
Average weight, 1930	+4.55* (24.32)	+.002 (0.155)	+.05* (5.07)	-.03*** (-1.69)	-.06* (-3.78)	.50
BMI, 1930	+3.63* (20.68)	-.01 (-.88)	+.03* (3.31)	-.01 (-.46)	-.06* (-3.78)	.44

Independent Variables

TABLE 22 continued

B. Absolute Gains (1926/1937) in Anthropometric Measures

Dependent Variable	Constant	Independent Variables					Adj. R^2
		Height, 1926 Levels	Weight, 1926 Levels	BMI, 1926 Levels	Paddy^c	Tenancy^d	
Gain in average height	+.54* (2.67)	-.33* (-2.60)	n.e.	n.e.	+.0003** (2.084)	-.0004*** (-1.985)	.17
Gain in average weight	+9.40* (2.82)	n.e.	-.18* (-2.72)	n.e.	-.006 (-.727)	+.02*** (1.86)	.28
Gain in BMI	.72 (.41)	n.e.	n.e.	-.04 (-.50)	-.01** (-2.50)	+.02* (3.03)	.22

NOTES: ^{a}Cross-sectional regressions on figures for the 30 prefectures with average percentage labor force in primary industry (PPI) for 1920, 1930, and 1940 greater than or equal to 55%.

^{b}Average values (for 1927 and 1935) of the 1927-based index of public health and medicine (see table 20) in the case of the regressions on % tall and % short. For the regressions on 1927 data the index is for 1927 and in the case of regressions on data for 1935 the index is for 1935.

^{c}The proportion of cultivable acres in the prefecture in paddy production multiplied by the average PPI value for 1920, 1930, and 1940 for the prefecture.

^{d}The proportion of cultivable acres in the prefecture held in tenancy multiplied by the average PPI value for 1920, 1930, and 1940 for the prefecture.

^{e}Nominal income per capita.
*Significant at the 1% level.
**Significant at the 5% level.
***Significant at the 10% level (all two-tailed tests).
t-statistics given in ( ) below parameter estimates.
n.e. = not entered into the regression.

1940 exceeds or is equal to 55 percent). I regress various measures of height for military recruitment examinees (i.e., percentage tall, percentage small, and average height) and the body mass index against the index of public health and medicine per capita, average real income per capita, and the percentages of cultivable acres in the prefecture (multiplied by the PPI variable to adjust for the degree to which the prefectural population was actually engaged in farming) farmed in tenancy, and the percentage of the cultivable acres in the prefecture (again multiplied by PPI) devoted to paddy rice production. As can be seen, tenancy is associated with lower levels of population quality; and in the regressions the elasticities on the tenancy variable are considerable. Moreover, as can be seen from panel B, tenancy was associated with slightly diminished gains in heights between 1926 and 1937 (and with slightly elevated gains in weights and BMIs). In short, lower levels of income were associated with lower levels of net nutritional intake and lower levels of population quality.[3]

It is also possible to use the estimates in panel B of table 22 to say something about long-run coevolution in rural Japan. Analysis of skeletal remains suggests that the population of Japan was on average taller before the Tokugawa period than during the Tokugawa period. G. Honda and T. Shay (1994: 13) give the following estimates on mean heights (in cm) for males and females in different periods of Japanese history (it goes without saying that the sample sizes on which the estimates are based are very small):

Period	Dates	Males (cm)	Females (cm)
Jōmon	600–200 B.C.	148.0	159.1
Yayoi	200 B.C.–A.D. 250	150.5	161.4
Kofun/Nara/Heian	250–1185	151.5	163.1
Muromachi/ Momoyama	1333–1600	146.6	156.8
Tokugawa	1600–1868	145.6	157.1

Could the diffusion of rice cultivation during the Tokugawa period have led to a diminution of height due to some long-run coevolutionary adjustment, say, to diet or to working in paddy rice fields? The estimates in table 22—albeit for the twentieth century and, therefore, perhaps of limited relevance for the preindustrial period—do not lend credence to this thesis.

To summarize: after the government opened up Japan to rice imports following the Rice Riots of 1918, marginal farmers and tenant farmers, concerned about a potential decline in relative income and greater instability in income-generating opportunities, brought on by the governmental tilt toward the industrial sector and by the new market conditions associated with unbalanced economic growth, suffered growing anxiety and frustration. This pent-up frustration exploded during the interwar period. And as had the ikki of the late Tokugawa period, it sent a signal to the central authorities in Tokyo to address the plight of the agricultural sector through the creation of a new entitlement system to replace the now-defunct Tokugawa entitlement system. In demanding changes the farmers were expressing concern about more than their own lives. They were also expressing concerns about their children and the long-run viability of their ie lines because of the feedback of labor productivity onto future population quality and future labor productivity that existed in rural Japan. Moreover, the population quality of districts characterized by high levels of tenancy was declining relative to that in the rest of rural Japan. And, since rural districts were declining in relative terms compared to industrial districts, the population quality of areas with high levels of tenancy was declining all the more relative to nonagricultural Japan.

INDUSTRIAL REGIONS, INDUSTRIAL OCCUPATIONS, AND ANTHROPOMETRIC MEASURES

In the face of the rural crisis, the government continued to pursue reforms aimed at dealing with the breakdown in health-enhancing labor market institutions within the industrial sector. A variety of factors explain the government's intense concentration on promoting reforms in the industrial sector: the pressure of the International Labor Organization, which Japan joined; the momentum gained by the reform-minded liberal wing of the bureaucracy from implementation of the Factory Act in 1916; the growing interest of employers and professional managers in health-enhancing programs because of their potential payoff in terms of improved worker morale and productivity, which reduced resistance to reforms in the business community; and the spread of socialist ideas among a small but growing labor movement in industrial districts.[4] The Health Insurance Law passed the Diet in 1922, although it was not actually implemented until 1931, the same year a law providing aid to injured workers was passed. And in 1916 the Nōshōmushō (Ministry of

Agriculture and Commerce) issued directives listing a group of occupational diseases to which special attention was to be paid by factory and
mine inspectors (however, silicosis for miners was not included in the
original list and was not placed on the list under 1930). Perhaps the
most important single factor in this trend was the decline in resistance
to reforms in the business community, which seems to have emanated
from two sources: from the evidence generated from a variety of industrial health studies that demonstrated that healthier factory workers
were more productive; and from the fact that as employees successfully
brought the labor boss system to an end, they found themselves increasingly devoting resources to the training of their blue-collar workers, and as their investment in these workers rose so did their interest in
keeping them through a host of paternalistic programs.[5] In short, the new
market environment was encouraging a positive attitude on the part of
the business community—especially large enterprises—to the creation
of a new system of health-enhancing entitlement programs, especially
those programs that linked entitlements to labor contracts, giving workers incentives to work hard and to remain with the company.

There was a convergence between market forces and governmental
(and military) interest in enhancing population quality that was making
the task of building up a new system of health-enhancing entitlements
far easier in urban industrial districts than in agricultural regions. That
the children of urban households tended to do less physical work than
those in villages—the Factory Law of 1911 banned child labor—and
that per capita expenditures on public health and medicine were far
higher in urban areas than in less densely settled villages were certainly
factors. That these environmental factors were playing a role can be
seen from table 23, which reports on regressions for the sixteen nonagricultural prefectures of Japan. The dependent variables are, respectively, gains in average height, gains in average weight, and gains in the
body mass index for recipients of military recruitment examinations between 1926 and 1937. In short, urban employers benefited from the environments they operated in.

But the improvement of conditions in industrial zones was not simply due to a better environment. As factory owners became more aware
of the potential benefits to productivity of reducing pollution on shop
floors and as factory inspectors pointed out substandard levels of ventilation and grime, and so forth, and assessed fines, industrial work conditions improved. An excellent test of the hypothesis that factory improvements were important is afforded by a comparison between the

TABLE 23

Factors Underlying Changes in Anthropometric Measures for Males Receiving
Military Recruitment Examinations in the Sixteen Industrial Prefectures, 1926-1937
(Nonlog-Log Regressions)

	Constant	Initial Level for Height, etc.[b]	Change in Income per Capita	Change in Index of Public Health/Medicine	Change in PPI[c]	Adjusted R^2
Change in average height[a]	.30 (1.32)	-.18 (-1.29)	.0000008 (.50)	.00005*** (1.75)	-.00008 (-.32)	.16
Change in average weight[a]	17.94* (4.40)	-.32* (-4.17)	-.000005 (-.70)	.01 (1.50)	-.09* (-4.42)	.75
Change in average BMI[a]	6.54* (4.12)	-.31* (-4.01)	-.000004 (-1.31)	.001 (.54)	-.03* (-3.51)	.72

NOTES:
[a]1937 value of variable minus 1926 value of variable, except for income per capita where it is the 1930 value of the variable minus the 1920 value of the variable.

[b]Initial level of average height in the case of the regression on change in height; initial level of average weight in the case of the regression on the change in weight; initial level of BMI in the case of the regression on the change in the BMI.

[c]Percentage labor force (both sexes) in primary industry (agriculture and forestry).

*Significant at the 1% level (two-tailed test).

**Significant at the 5% level (two-tailed test).

***Significant at the 10% level (two-tailed test).

relative height and weight of female textile workers around 1910 with the relative height, weight, and BMI of textile factory workers during the late 1930s. By "relative," I mean relative to students. Consider the figures in panel C of table 24 which we can usefully compare to those in table 18. In making the comparison it is important to keep in mind that female textile workers tended to be drawn from tenant farm households, that is, from households whose relative levels of population quality were in decline. And yet as can be seen from table 25 by simply taking arithmetic differences between the figures for the interwar period and for 1910, the gap in human growth measures between female students and factory girls was decisively closing. More nutritious meals in company cafeterias, regular periods of rest, improved ventilation, and more attentive medical staffs all made a contribution. By late interwar Japan, the large-scale factory—that covered by the Factory Act—was no longer an impediment to the convergence in population quality differentials as it had been during the era of balanced economic growth.

The other tabular material in table 24 is also of interest. The figures in panel A show that the BMI tends to be smallest at the extremes of the height spectrum and that it is largest somewhere in the middle of that spectrum. And the figures in panels C and D show that food intake was greatest for groups whose physical activity was the most demanding; but it also shows that in spite of the greater gross nutritional intake of those doing heavy physical work, heights and weights tended to be greater for those consuming less food and subjecting their bodies to lower levels of physical exertion. In short, these data underline the importance of demands placed on nutrient intake.

THE EMERGENCE OF GOVERNMENT
WELFARE POLICY IN INTERWAR JAPAN

During the interwar period, as unbalanced economic growth transformed labor, capital, and product markets, the government found itself increasingly drawn into the problem of building a new system of entitlements to replace the old balkanized Tokugawa system fallen into disuse. Political pressure from workers and especially from tenant unions and marginal farmers played a fundamental role in bringing the government's attention to this issue. And by the 1920s the liberal wing of the national bureaucracy was committed to introducing a program of reform that was inspired by that already introduced in many Western European countries and in the United States. Had the military crisis not

TABLE 24

Various Anthropometric Measures for Military Recruits, Students, Factory and White-Collar Workers, and Farmers in Interwar and Early Pacific War Japan

A. Males Age 20, BMI in Height Classes, 1933 (A Group Tallest, P Group Shortest)[a]

Group	A	B	C	D	E	F	G	H
BMI	19.8	20.0	20.1	20.2	20.3	20.5	20.6	20.6
Group	I	J	K	L	M	N	O	P
BMI	20.7	20.8	20.8	20.8	20.7	20.7	20.5	20.3

B. Height, Weight, and BMI of Male Students and Workers at Various Ages

Students				Miscellaneous Workers			
Age	Height (cm)	Weight (kg)	BMI	Ages	Height (cm)	Weight (kg)	BMI
12	137.1	31.3	16.7	17-18	159.9	51.0	20.0
13	142.5	35.0	17.2	19-20	160.9	53.9	20.8
14	148.5	40.2	18.2	21-25	161.2	50.9	19.6
15	154.9	45.0	18.8	26-30	157.9	52.5	21.1
16	159.7	48.0	18.8				
17	159.5	48.0	18.9				

TABLE 24 continued

C. Height, Weight, and BMI for Female Students and Workers

	Students				Workers, Mostly Spinning Mill Operatives			
Age	Height (cm)	Weight (kg)	BMI	Age(s)	Height (cm)	Weight (kg)	BMI
12	137.5	31.7	16.8	12	136.8	34.4	18.4
13	143.5	36.0	17.1	13	141.1	37.6	18.9
14	146.9	40.6	18.8	14	143.4	40.9	19.9
15	149.0	42.8	19.3	15	145.5	43.5	20.5
16	149.7	44.5	19.9	16	145.2	44.6	21.2
17	149.3	47.5	21.3	17	146.0	46.2	21.7
				18	146.3	47.9	22.4
				19-20	147.0	47.7	22.1
				21+	146.3	48.1	22.5

TABLE 24 continued

D. Nutrient Consumption and Summer Weight for Males in Various Occupations, Early 1940s

	Physical Laborers		Light Physical Laborers				Villagers	
	Heavy	Moderate	Educators	Police Admin.	Shopkeepers	Agricultural	Fishing	
Calories	3,919	3,027	n.a.	n.a.	n.a.	n.a.	n.a.	
Proteins	126.3	96.8	n.a.	n.a.	n.a.	n.a.	n.a.	
Weight, ages 13-16	n.a.	n.a.	45.3	n.a.	43.1	49.5	41.4	
Weight, ages 17-34	57.8	n.a.	53.7	59.2	50.8	53.0	54.9	
Weight, ages 35-55	56.2	n.a.	57.0	61.3	54.7	52.1	55.9	

E. Summer Weight for Females in Various Occupations, Early 1940s

	Physical Laborers[b]		Light Physical Laborers			Villagers	
	Heavy	Moderate	Nurses	Shopkeepers	Agricultural	Fishing	
Ages 13-16	44.9	44.9	48.6	37.5	41.2	37.9	
Ages 17-34	47.0	48.0	50.4	46.2	46.0	47.3	
Ages 35-55	n.a.	n.a.	41.0	47.4	42.4	49.1	

SOURCES: Nihon Naikaku Tokeikyoku 1933: 410-411; Teruoka 1942: various tables; Yagi 1935: various tables.

NOTES: [a]The heights for individuals in each height group are contained within a range of 2.5 cm. Group A, for instance, has individuals of heights between 177.5 and 179 cm, and at the other extreme, the heights in Group P are between 140 and 142.5 cm. To estimate the BMI for each group, I assumed that the average height was the average of the upper and lower heights in the group (the original table gives average weights for the height groups).

[b]Heavy physical workers are workers in silk filatures, and light physical workers are telephone operators.

n.a. = not available.

TABLE 25

Changes in Anthropometric Measures for Female Students and Workers, 1910/1930

Students

Age	Absolute Gains, 1930s/1910			Percent Gains, 1930s/1910		
	Height (cm)	Weight (kg)	BMI	Height (cm)	Weight (kg)	BMI
15	6.0	4.2	0.4	4.2	10.9	2.1
16	3.3	2.1	0.1	2.3	5.0	0.6
17	2.0	2.5	0.5	1.4	5.6	2.4

Factory Operatives

Age	Absolute Gains, 1930s/1910			Percent Gains, 1930s/1910		
	Height (cm)	Weight (kg)	BMI	Height (cm)	Weight (kg)	BMI
15	8.8	7.8	1.4	6.4	21.9	7.3
16	5.8	5.6	1.1	4.2	14.4	5.5
17	5.1	4.1	0.5	3.6	9.7	2.4
18	4.8	3.5	0.2	3.4	7.9	0.9

grown over the course of the interwar period, that program, or at least key parts of it, might have been instituted. As it was, a new centralized health-enhancing entitlement system did emerge: the key features of that program were the Factory Act, health and disability insurance for industrial workers, and, in 1938, the creation of the National Health Insurance Law which aimed at creating, with central government assistance, a system of health insurance societies in local communities, especially in agricultural districts.

But the military crisis grew worse. And as it did so, the military found itself caught on the horns of a terrible dilemma. Military spending diverted scarce tax resources and bureaucratic attention and energy away from population quality-enhancing social welfare programs. And yet the military, perhaps more than any other central government agency, was concerned about the perpetuation and even widening of differentials in height and weight and body mass index.

After Japan's surrender, freed from the burden of making excessive military expenditures and from the ideological trammels of fascism imposed by the conservative wing of the bureaucracy (many members of which were purged after 1945), the liberal wing of the bureaucracy was able to decisively complete the program of reform envisioned in the 1920s and 1930s. With bewildering speed the government passed a tough labor standards law governing work conditions and labor contracts, sweeping land reform that did away with absentee landlordism, price supports for rice that guaranteed a minimum standard of living for marginal farm households comparable to that enjoyed by blue-collar households, and a comprehensive national health insurance law. By the early 1960s Japan had joined the ranks of the advanced industrial nations in terms of its system of national entitlement programs.

But passage of this sweeping program of reform, realizing the dreams of the liberal bureaucrats of the 1920s, had to wait until the late 1940s, the 1950s, and the early 1960s. It was hardly uppermost in the minds of the Japanese bureaucrats—it was certainly not uppermost in the minds of the military planners—during the late summer of 1941 as Japan plunged into full-scale war against its great Pacific rival, against the very country that had forcibly opened it up to international commerce during the 1850s, the United States.

CHAPTER 6

Conclusion

Within a period of twelve decades after its feudal government collapsed in ruins, Japan, a country by and large lacking natural resources and geographically isolated from the Atlantic region that dominated international trade before World War I, achieved remarkable growth in per capita income through industrialization, growth that reached the level obtaining in some Western European nations. It takes no great leap of imagination to conclude that Japan managed to catch up with the West because she was able to develop her human resources. Indeed, the improvement in her standard of living, as Sen defines it, in terms of capabilities, certainly serves both as the best explanation for Japan's successful growth and the best measure of that growth. Surely measuring Japan's standard of living in terms of capabilities is superior to measuring it in terms of opulence precisely because the capabilities definition includes both a yardstick for gauging the level and speed of growth and an explanation for growth. However, practical measurement of the standard of living defined in terms of capabilities is an exceedingly difficult, perhaps hopelessly impossible, task. For this reason I propose a substitute measure: population quality measured in terms of standard anthropometric variables like height, weight, chest girth, and the body mass index. And because I am interested in a measure of population quality that comes close to capturing capabilities, I focus on the indicators of human growth for children and young adults.

Using data on schoolchildren it is possible to construct for Japan an annual auxological data set for children aged 6, 12, and 18 that extends from 1900 to 1985 and that gives us a clear picture of a remarkable secular improvement in population quality. Combining these data with estimates of net nutritional intake it is possible to show that the main factor underlying the great improvement in population quality is the secular trend in net nutritional intake. That the population quality of Japan is far higher today than it was in the 1870s is because of improvements in agricultural technology that raised per capita levels of gross nutritional intake; because of greater levels of expenditure on public health and medicine and an awareness on the part of doctors and public health officials of the germ theory of disease and antibiotic drugs; and because children in Japan now spend their formative years in school rather than working under the blistering sun in the rice fields. Thus one can say that the great secular improvement in population quality is the end result of the triumphant spread of Western scientific knowledge and of industrialization to a remote island nation off the Chinese mainland.

One can certainly summarize the central message of my statistical analysis in this way. But to do so is, in my opinion, a great mistake. For the statement does not do justice to the social struggle Japan went through as it transformed its system of health-enhancing entitlements from the balkanized form characteristic of feudalism to the modern form in which programs emanate from a centralized national government in Tokyo. Demand matters; or rather the way demand is voiced through markets and through the political process matters.

No better proof of the importance of demand forces can be offered than the account that constitutes Part II of this volume, an account that details the gradual breakdown in health-enhancing market institutions and in the balkanized system of entitlements developed during the late feudal period. The breakdown stretched over a period of almost half a century, in part because the labor market structure and the institutions governing contracts between employers and employees which developed during the late feudal period persisted into the balanced economic growth era following the demise of feudalism, in part because the balkanized system of entitlements had become deeply rooted in Japan's society and polity, especially in landlord-tenant relations. Indeed, Japan's ability to successfully absorb Western technology in a short time must be partly attributed to the development of population quality during the late feudal period, a development that was encouraged by market forces and by the entitlement system.

Nonetheless, the feudal system eventually did break down. These institutions were no longer viable in a world of large industrial enterprises and in a world in which harvest crises rarely, if ever, occurred. But what was to replace them? The liberal bureaucrats had a vision of what could and should replace these institutions just as surely as industrialists had a vision of what machinery should replace traditional Japanese silk reeling equipment: the entitlement systems developed, and evolving, in Germany, France, and the United States. But to have a vision and to realize it is another matter, and the liberal bureaucrats faced two major obstacles to realizing their vision. One was represented by remnants of the balkanized system of feudal entitlements, especially that part of the Tokugawa system embodied in landlord-tenant relations. And the other was militarization and the growth of fascist ideology among some segments of the bureaucracy. The legacy of the past acted as a barrier; so did Japan's military expansion into the rest of Asia and her alliance with fascist Germany. Despite widespread rural unrest and the diffusion of socialist ideology in a nascent industrial labor movement—the two movements vociferously crying out for reform of entitlements and a restructuring of market institutions—the impediments to the development of a sweeping program of entitlements proved so great that reform remained anemic at best. The hallmarks of that reform—the Factory Act of 1911 and the health insurance legislation of 1938—were hardly earthshaking. At the root of the problem was the program of militarization Japan was increasingly committed to after the late 1920s. Hence the vision of the liberal bureaucrats was not realized in the form of comprehensive legislation until the late 1950s and early 1960s when Japan had in place a comprehensive land reform, a rice price support program that guaranteed a minimum standard of living for marginal farmers, a strict labor standards act, and a national program of health insurance and old age pensions.

But while the program of reform was not achieved until militarization of Japanese society ended and the fascist bureaucrats had been purged from the ministries in Tokyo, that program of reform echoed the demands of the tenant and labor unions of the interwar era when Japan decisively entered into a period of unbalanced economic growth. Those who dedicated their efforts to these causes did not do so in vain. My focus on demand in the development of Japan's population quality is testimony to their sacrifices.

To return by way of closing to the metaphor with which I began this account: in comparison to their ancestors, the people of Japan today are giants on the face of the earth. They are giants today because they command a more powerful technology than they did in the past; they are giants today because they have moved from a rice-producing agrarian nation to an industrial power; and they are giants today because the ordinary people of Japan demanded it and voiced that demand in the streets of the cities, on the shop floors of the factories, and in the paths of the remote villages leading up along the steep slopes of the mountains.

Notes

CHAPTER 1. INTRODUCTION

1. The quote is from Genesis 6:4 (Cambridge: Cambridge University Press, n.d.: New Testament, p. 7). Floud, Wachter, and Gregory (1990) also make use of this quotation at the beginning of their volume but draw different implications from it. On the methodological importance of analogy and metaphor in the sciences and social sciences, see inter alia, Nelson, Megill, and McCloskey 1987; McCloskey 1994; Mirowski 1987; and the discussion in chapter 2 ("Long-Term Determinants of the Anthropometric Measures, 1901–1979," and appendix) of this volume.

2. The relationship between the standard of living and what I call population quality is a thorny one. I distinguish between the standard of living and population quality on practical grounds. In common parlance indicators of potential physical strength and endurance and of intelligence and learning capacity reside within the organs and tissues of the individuals within a population and hence tell us something about the population's average quality. The standard of living is either a more subjective matter involving a psychological sense of self-worth and satisfaction or a matter of goods and services consumed and possessed by individuals within a population and hence external to physical bodies. As I argue in the text of this section and the next section of chapter 1, there is a relationship between Sen's definition of the standard of living, which revolves around capabilities conceived in terms of the physical and psychological capabilities to take advantage of economic and social opportunities, and my definition of population quality in the sense that both of us reject definitions based on per capita consumption of goods and services or on utility. However, my definition is narrower than his since I exclude a host of political and social factors like political freedom which Sen considers important.

Another reason I make a point of distinguishing between population quality and the standard of living defined in terms of per capita consumption is that I feel some scholars mistake the one for the other, or use the former as a proxy for the latter. For instance, in Fogel (1986) and in many of the fine papers in the volume edited by Komlos (1994) there is a tendency to confuse anthropometric measures with the standard of living. As an example, consider the title of Shay 1994. While I do think some aspects of what these scholars define as the standard of living, especially food consumption, are important to determining population quality as I define it, I do not believe that population quality is equivalent to these components of the standard of living defined in consumption terms. Moreover, I favor Sen's definition of the standard of living over that defined in terms of consumption of goods and services because I wish to emphasize the linkage between the standard of living and work capacity.

Steckel (1994a, 1994b) reviews some other proposed indexes that bear a resemblance to what I propose here: for instance, a physical quality of life index advanced by Morris and the UN development index and his own biological standard of living index. However, none of these is identical to population quality. To some extent, of course, it does not matter what we call the underlying variable we are measuring with the anthropometric proxy variables as long as we are clear about what we think statistical analysis is actually telling us. For instance, in this volume I utilize anthropometric measures as proxies for population quality, but persons who prefer to interpret results concerning the determinants of the anthropometric measures in terms of another conceptual framework in which the anthropometric measures are proxies for something else are free to do so. I do attempt, however, to show in Part II that an interpretation in terms of capabilities and work capacity is reasonable.

3. The influence of the gene pool on anthropometric measures is one reason that it is difficult to make inferences about the standard of living of slaves in eighteenth- and nineteenth-century America on the basis of comparisons between figures on height and weight for slaves and nonslave populations like Irish immigrants to Boston. Of course, if we adopt Sen's concept of the standard of living, slavery, by constraining human freedom, automatically reduces the standard of living of the enslaved population.

4. There is a literature in contemporary Japan known as the Nihonjinron which argues that the Japanese people constitute a racially distinct people. It is sometimes argued by advocates of this viewpoint that the digestive organs of Japanese differ from those of non-Japanese or the Japanese people learn the Japanese language on one side of the brain while other racial groups learn their languages on the other side of the brain, and so forth. Miller (1982) reviews many of these arguments, demonstrating that the experimental evidence adduced in their favor is shaky or nonexistent. He also shows that many of the arguments, such as that concerning language and the brain, are circular. The obvious problem with this is that since many individuals not of Japanese origin, like Koreans living in Japan whose mother tongue is Japanese, learn Japanese, and they either learn the language on the other side of the brain from the Japanese—in which case they are not really learning Japanese as a language—or they learn the language on the same side of the brain as the Japanese—in which case by

definition they are themselves Japanese even though they are not Japanese. It should be obvious from the text that I think the idea of a pure Japanese "race" is not a useful concept and I reject it throughout this book.

5. There is a large literature that deals with the potential interactions between the environments phenotypes function in and genetic evolution. One line of argument is sociobiological and posits that maximizing survival of the gene pool may be the basis for altruistic behavior toward one's own children and kin. On this literature, see Goldsmith 1991 and Wallace 1972a, 1972b.

6. Our knowledge of the specific impact that the nutrients in food (and their processing through cooking and preparation for eating) have on human growth and potential physical and mental exertion and to long-run coevolutionary change over time certainly leaves much to be desired. Nevertheless, progress has been made. See, for example, Cohen 1987; Durnin 1983; Ebrahim 1979; Eide and Steady 1980; Gibson 1990; Greksa, Pelletier, and Gage 1986; Harris and Ross 1987; Jelliffe and Jelliffe 1979; Pellett 1987; Pike and Brown 1967; Taylor 1982.

CHAPTER 2. SECULAR TRENDS IN ANTHROPOMETRIC MEASURES OF HUMAN GROWTH

1. Unless otherwise indicated, the term "height" refers to standing height. Later on in this chapter we shall encounter figures on sitting height, and I will specifically use the term "sitting height" when this is what I am discussing.

2. Shay (1994) relies on figures for military recruits for his analysis. This makes a good deal of sense given the fact that he concentrates on regional analysis rather than time series analysis (the data for recruits are available for regions during the early twentieth century and later on for prefectures). In chapters 4 and 5 I provide a much less systematic analysis of regional differentials in height and the other anthropometric measures for military recruits than does Shay and therefore recommend that the reader interested in regional differentials consult his interesting analysis.

3. I am grateful to Professor Osamu Saito of the Institute of Economic Research at Hitotsubashi University for pointing this out to me.

4. For the weights and for comparisons between calorie intake on a per capita and on a per consumer unit weighted basis, see table 1.

5. Kurosaki (1967) provides an excellent discussion of regional variation in food intake during the preindustrial period. He shows that in some areas of southern and northern Japan, for example, in southern Kyushu, potatoes, not rice, were the main staple foodstuff. Kito (1986) also gives figures on food consumption for regions of Japan during the nineteenth century.

6. After the Rice Riots of 1918 Japan began to systematically import rice and other foodstuffs from its colonies, Korea and Taiwan. The figures on food consumption are based on consumption and not production and therefore take account of imports. For a discussion of imports and food policy before World War II, see Brandt 1993. For a discussion of Japanese food policy during World War II, see Johnston 1953. For a discussion of imports after World War II and estimates of income and price elasticities of demand for various foodstuffs, see Sanderson 1978. Uchino (1977) provides an excellent treatment, based on sur-

veys of households, of the changing nature of diet and of the role of urbanization in the shift toward more "Western" types of food consumption.

CHAPTER 3. THE TOKUGAWA LEGACY

1. The entitlement concept used in this way is from A. K. Sen. For his use of
the notion, see Sen 1987 and Sen et al. 1987. For a discussion of political unrest
and the struggles over entitlements in European history, see Tilly 1983.

2. My treatment of the establishment and operation of the Tokugawa political system and the consequent expansion in rice cultivation is cursory. For
much more detailed treatments, see Beasley 1972; Kikuchi 1986; Ramseyer
1989; Smith 1988; Totman 1967; and Vlastos 1986.

3. I am grateful to Hiroshi Kawaguchi of Tezukayama University for explaining this point to me. For a monumental treatment of the development of
shinden in Japan, see Kikuchi 1986.

4. For a discussion of urbanization during the late Tokugawa period and the
expansion of Nagoya in particular, see Mosk 1995b.

5. An interesting question whose full discussion would carry us far beyond
the thematic concerns of this volume is whether the stem family system came
into existence at the end of the early Tokugawa period *because* population had
been growing rapidly over the preceding century. That is, after a century of substantial population growth Japan was reaching a point at which constraints on
new land availability characterized by a highly inelastic supply curve for newly
developed arable land was limiting the ability of families to divide land or find
land for sons and daughters in new villages or to arrange for successful marriages through which sons or daughters leaving the family's estate could expect
to enjoy a reasonable level of sustenance. Some scholars, like Smith (1959,
1988), argue that an inelastic supply of new arable land and technological
changes led to the breakup of extended families into smaller stem family units,
that is, that the stem family was basically a by-product of the new economic environment at the end of the early Tokugawa period, in particular, a by-product
of the problem of inheritance in an environment in which land could not easily
be divided any further. Other scholars argue that the statistically observed decline in family size whereby small families emerged in the Tokugawa period
after a time when large family units were recorded in the temple registers is
largely an artifact due to changes in registration practices (see Mosk 1995a for
a brief discussion of this issue). In any event, regardless of which viewpoint one
adopts, one cannot deny the independent role of culture in creating the stem
family system in Japan. The reason is, I think, obvious: there are many societies
where new arable land is inelastically supplied and yet they do not develop a
stem family system.

For a discussion of the post-1990 diffusion of the samurai model of the stem
family system into middle-class and farming populations and the role of the educational system in this diffusion process, see Smith 1983.

6. For a theoretical discussion of this problem and empirical evidence concerning the relationship between infant mortality and ages for first marriage for
women in Japan, see Mosk 1983.

7. That families in Japan used in-adoption of fictive heirs in such instances raises an interesting question as to whether such families could be said to be maximizing reproductive success in terms of the gene pool, an idea put forward by sociobiologists. The issue is conceptually a thorny one, and I raise it here simply as one to consider.

8. For recent articles on the practice of infanticide and mortality in Tokugawa Japan, see Cornell 1994; Janetta and Preston 1991; Morris and Smith 1985; Ohta and Sawayama 1994; Saito 1993; Smith 1988; Tomobe 1994. Kawaguchi (1993) gives graphic details from the diary of a household concerning infanticide. For a discussion of the quality of the shumon-aratame-chō registers, see Cornell and Hayami 1986.

9. That the fiefs provided this entitlement insurance is one reason in my opinion why peasants were encouraged to reduce their fertility. The problem peasants face in many agricultural settings is a highly variable probability of adequate harvest. In such a risky setting households are likely to pursue a "safety first" strategy, which means not taking excessive risks in crops planted and avoiding small family sizes, since infants are likely to die in times of dearth. But because of entitlement insurance, even households in regions where harvests were often poor did not tend to pursue a "safety first" strategy as far as fertility was concerned.

CHAPTER 4. POPULATION QUALITY IN AN ERA OF BALANCED ECONOMIC GROWTH, 1880–1920

1. For a treatment of labor markets during the Meiji period which discusses the wage floor issue, the existence of surplus labor in agriculture (i.e., the existence of a labor pool in which the marginal product of an hour's worth of work is positive but in which workers can be removed without loss of output because the remaining workers work harder), see Mosk (n.d.: chap. 2). Mosk (1995c) also discusses the issue of how nonwage payments (like subsidized housing and meals) affect measured wage levels and wage differentials between agriculture and nonagriculture. In principle, none of the wage figures given in table 16 includes the market value of nonwage remuneration. For a discussion of the development of training programs within companies, see Iwauchi and Sasaki 1987.

2. Most recruits to the textile mills came from tenant households, and the contracts were typically signed by the head of the household on behalf of the girl involved, a fact demonstrated by Tsurumi (1990). Hence the relevant opportunity cost for the agricultural sector for a female textile worker was the marginal productivity of her labor on tenant farms. I deal with productivity and income differentials between owner, part-owner/part-tenant, and tenant farm households in chapter 5. On the importance of female labor inputs in rural Japan during the late feudal and early modern periods, see Saito 1991.

3. For further and more detailed discussion of these issues, see Gordon 1985 and Mosk 1995c.

4. My discussion concerning the importing of Western medicine and public health during the late Tokugawa and Meiji periods draws on the accounts in Fukutome 1986; Hirota 1957; Miura 1978, 1980a, 1986; Nihon Kōseishō 1955; Sugaya 1976, 1982; and Sugita 1969.

5. A highly restrictive set of relief measures known as *jutsukyū* (or *jukkyū*) *kisoku* (relief regulations) were carried over the bakufu domain and some of the fiefs. But this hardly constituted a program of entitlements since it was highly selective in who qualified for coverage. The law had special provisions for those over seventy and for those under thirteen years of age who did not have an adult taking care of them, and otherwise it emphasized the need for self-reliance. For a discussion of the slow development of entitlement policy during the Meiji period and later, see Adachi 1988; Fukutome 1986; Hotani 1994; and Onishi 1988. Yamanaka (1966) provides a useful history of the development of private and public initiatives in the life insurance industry.

6. I classify the prefectures by levels of urbanization here (figures for the prefectures on labor force structure and per capita income are not available for this period), but I wish the reader to be aware that there is a high correlation between levels of urbanization, the proportion of persons employed outside of primary industry, and per capita income. On this point please refer to table 20 and to the discussion surrounding table 20 in chapter 5 below.

7. Emi (1963) provides a useful account of national spending on social security and other programs.

8. Miura (1980a, 1980b) gives details concerning the various governmental and nongovernmental research agencies set up to study conditions in the factories during the Meiji and Taisho (1915–1926) periods.

9. Honda and Shay (1994) demonstrate that throughout the period 1899–1937 average heights of military recruits in agricultural districts (defined as those prefectures with 60 percent of its labor force in primary production in the year 1930) fell consistently below the average for the industrial prefectures (those with less than 40 percent of its labor force in primary production in 1930) and the absolute level of the difference in average heights was actually increasing somewhat throughout the 1899–1937 period. Shay (1994), using the same prefectural data and making comparisons in the ranks of the prefectures according to average height for military recruits for the years 1898, 1902, 1907, 1912, 1917, 1922, 1927, 1932, and 1937, shows that while prefectures producing on average short and tall recruits tended to remain the same, the ranks of the prefectures within the groups of prefectures producing tall and short recruits did vary over time. One problem with both of these analyses is the use of average male height figures: the absolute differences in averages between the prefectures at adulthood tends to be very small. In this context see table 20 in chapter 5.

For an illuminating study of the changing regional pattern of illiteracy in Meiji Japan, which bears on the question about continuity and discontinuity in regional patterns during the era of balanced growth, see Kurosaki 1983.

CHAPTER 5. ENTERPRISE, COMMUNITY, AND HUMAN
GROWTH DURING AN ERA OF UNBALANCED ECONOMIC
GROWTH, 1920–1940

1. For a general history of tenant unions and landlord-tenant disputes in Japan, see Smethurst 1986 and Waswo 1977, 1982.

2. Smethurst (1986) argues that rebellion broke out in rural Japan because tenant farmers who had been enjoying an improvement in income for some time

wanted to improve even more their commercial market opportunities by doing away with the practice of paying rents in the form of rice, thereby allowing diversification into a wider range of crops, and because tenants paid excessively high rents (typically about 40 percent of the tenant's crop). At one time high rents might have been justified because of the entitlement insurance provided by landlords, but the demand for such "insurance" had been on the decline for a considerable time because of the secular improvement in productivity per acre. As appealing as the argument may seem, it has been harshly criticized on factual grounds. See in this regard the systematic critique by Nishida (1986). In my view the Smethurst thesis misses the point for two reasons: it focuses on absolute levels of income instead of relative levels of income; and it fails to take into account the fact that many farmers were seeking greater price and income stability through government intervention in the market aimed at guaranteeing minimum levels of income for marginal producers and guaranteeing tenant rights to continue to cultivate land they had been cultivating for many years. Stability of rights to continue to farm land became increasingly important as farm households, responding to the changing terms of trade and the rising relative cost of labor to land, increased their investments in machinery and other forms of capital equipment. For further discussion of Smethurst 1986, see Brandt 1989.

3. In tabulations that are not displayed in table 21, I found that small farmers tended to work longer hours per worker than large farmers, and other things being equal, part-owner/part-tenant households worked longer hours per worker than did owner households. And of the three groups I found that tenant households worked the longest hours per worker. From the last line of panel A of table 21 it is apparent that dependence on hired labor declined for all size and tenancy groups so that the demands of agricultural work were increasingly being met by family members regardless of ownership status. However, in tabulations not reported here, I found that owners used more hired labor than did part-owner/part-tenant households and that tenant households were the least likely to use hired labor. This is one reason tenant household members worked longer hours than did those in part-owner/part-tenant households or in owner households.

4. For further details, see Gordon 1985 and Mosk 1995c.

5. For a review of the various studies concerning industrial health conducted by governmental agencies and by private or semiprivate research institutes during the interwar period, see Kano 1974 and Miura 1980b, 1981.

References

Adachi, M., ed.
  1988   *Fukushi kokka no rekishi to tenbo* (The History and Prospects
              for Welfare Nationalization). Kyoto: Horitsubunkasha.
Beasley, W. G.
  1972   *The Meiji Restoration.* Stanford: Stanford University Press.
Brandt, L.
  1989   Review of R. Smethurst, *Agricultural Development and Ten-*
              ancy Disputes in Japan, 1870–1940. Journal of Economic His-
              tory 49 (3): 749–750.
  1993   "Interwar Japanese Agriculture: Revisionist Views on the Im-
              pact of the Colonial Rice Policy and the Labor-Surplus Thesis."
              Explorations in Economic History 30 (3): 259–293. Cambridge
              at the University Press.
  n.d.   *The Holy Bible, Authorized King James Version.* Cambridge:
              Cambridge University Press.
Cohen, M. N.
  1987   "The Significance of Long-Term Changes in Human Diet and
              Food Economy." In *Food and Evolution: Toward a Theory of*
              Human Food Habits, ed. M. Harris and E. B. Ross, 262–284.
              Philadelphia: Temple University Press.
Cornell, L. L.
  1994   "Infanticide and the Origin of Low Fertility in Early Modern
              Japan: A Sociological History." Paper presented to the Work-
              shop on Abortion, Infanticide, and Neglect in the Asian Past:
              Japan in East Asian Perspective, 20–21 October, International
              Center for Japanese Studies, Kyoto.

Cornell, L. L., and A. Hayami
    1986   "The *Shumon Aratame Cho:* Japan's Population Registers."
                Journal of Family History 11 (4): 311–328.
Crafts, N.
    1983   "Gross National Product in Europe, 1870–1910: Some New Es-
                timates." *Explorations in Economic History* 20: 311–401.
Denton, F.
    1988   "The Significance of Significance: Rhetorical Aspects of Statisti-
                cal Hypothesis Testing in Economics." In *The Consequences of
                Economic Rhetoric,* ed. A. Klamer, D. N. McCloskey, and R.
                M. Solow, 163–183. Cambridge: Cambridge University Press.
Durham, W.
    1991   *Coevolution: Genes, Culture and Human Diversity.* Stanford:
                Stanford University Press.
Durnin, J. V. G. A.
    1983   "The Variability of Dietary Energy." In *Energy Balance in Human
                Nutrition,* ed. J. Kevany, 13–23. Dublin: Royal Irish Academy.
Ebrahim, G. J.
    1979   "The Problems of Undernutrition." In *Nutrition and Dis-
                ease,* ed. R. J. Jarrett, 13–130. Baltimore: University Park
                Press.
Eide, W. B., and F. C. Steady
    1980   "Individual and Social Energy Flows: Bridging Nutritional and
                Anthropological Thinking about Women's Work in Rural
                Africa: Some Theoretical Considerations." In *Nutritional An-
                thropology: Contemporary Approaches to Diet and Culture,* ed.
                N. W. Jerome, R. F. Kandel, and G. H. Pelto, 85–116. New
                York: Redgrave Publishing Company.
Emi, K.
    1963   *Government Fiscal Activity and Economic Growth in Japan:
                1868–1960.* Tokyo: Kinokuniya Book Store.
Eveleth, P., and J. M. Tanner
    1990   *Worldwide Variation in Human Growth.* Cambridge: Cam-
                bridge University Press.
Floud, R., K. Wachter, and A. Gregory
    1990   *Height, Health and History: Nutritional Status in the United
                Kingdom, 1750–1980.* Cambridge: Cambridge University Press.
Fogel, R.
    1986   "Physical Growth as a Measure of Economic Well-Being of Pop-
                ulations: The Eighteenth and Nineteenth Centuries." In *Human
                Growth: A Comprehensive Treatise,* ed. F. Falkner and J. M.
                Tanner, 263–281. New York and London: Plenum Press.
Fukutome, S.
    1986   "Meiji senki ni okeru Nihon seifu no rōdōsha no rōdō eisei ni
                taisuru taido ni tsuite" (The Attitude of the Japanese Govern-
                ment Toward Industrial Health in the Early Part of the Meiji Pe-
                riod). *Rōdō* (Journal of Science of Labor) 62 (2): 77–92.

Gibson, R. S.
1990   *Principles of Nutritional Assessment.* Oxford: Oxford University Press.
Goldsmith, T.
1991   *The Biological Roots of Human Nature: Forging Links between Evolution and Behavior.* New York and Oxford: Oxford University Press.
Gordon, A.
1985   *The Evolution of Labor Relations in Japan: Heavy Industry, 1853–1955.* Cambridge: Harvard University Press.
Greksa, L. P., D. L. Pelletier, and T. B. Gage
1986   "Work in Contemporary and Traditional Samoa." In *The Changing Samoans: Behavior and Health in Transition,* ed. P. T. Baker, J. M. Hanna, and T. S. Baker, 297–326. New York: Oxford University Press.
Hanley, S.
1983   "A High Standard of Living in Nineteenth-Century Japan, Fact or Fantasy." *Journal of Economic History* 46 (1): 183–192.
Hanley, S., and K. Yamamura
1977   *Economic and Demographic Change in Preindustrial Japan, 1600–1868.* Princeton: Princeton University Press.
Harris, M., and E. B. Ross, eds.
1987   *Food and Evolution: Toward a Theory of Human Food Habits.* Philadelphia: Temple University Press.
Hayami, A.
1983   "The Myth of Primogeniture and Impartible Inheritance in Tokugawa Japan." *Journal of Family History* 8 (1): 3–29.
Hayami, Y. [In association with M. Akino, M. Shintani, and S. Yamada]
1975   *A Century of Agricultural Growth in Japan: Its Relevance to Asian Development.* Minneapolis: University of Minnesota Press.
Hirota, H.
1957   *Nihon igaku 100-nenshi* (100-Year History of Japanese Medicine). Tokyo: Rinsho Igakusha.
Honda, G., and T. Shay
1994   "Differential Structure, Differential Health: Industrialization in Japan, 1868–1946." Paper presented to the National Bureau of Economic Research Pre-Conference, Health and Welfare during Industrialization, July.
Hotani, R.
1994   *Nihon no shakai seisakushi* (The History of Japan's Social Policy). Tokyo: Chuo Keizaisha.
Iwauchi, R., and S. Sasaki
1987   "Industrialization and In-Company Training." In *Vocational Education in the Industrialization of Japan,* ed. T. Toyoda, 187–215. Tokyo: United Nations University.

Janetta, A. B., and S. H. Preston
1991   "Two Centuries of Mortality Change in Central Japan: The Evidence from a Temple Death Register." *Population Studies* 45: 417–436.
Japan Office of the Prime Minister, Statistical Bureau
various   *Japan Statistical Yearbook.* Tokyo: Office of the Prime Minister.
years
Japan Statistical Association
1987   *Historical Statistics of Japan.* Vol. 1. Tokyo: Japan Statistical Association.
1988   *Historical Statistics of Japan.* Vol. 5. Tokyo: Japan Statistical Association.
Jelliffe, D. B., and E. F. P. Jelliffe, eds.
1979   *Nutrition and Growth.* New York: Plenum Press.
Johnston, B. F.
1953   *Japanese Food Management in World War II.* Stanford: Stanford University Press.
Kagoyama, T.
1970   *Jokō to kekkaku* (Female Factory Workers and Tuberculosis). Tokyo: Koseikan.
Kalland, A., and I. Pedersen
1984   "Famine and Population in Fukuoka Domain During the Tokugawa Period." *Journal of Japanese Studies* 10 (1): 31–72.
Kawaguchi, H.
1993   *18–19 seiki ni okeru Aizu-Minamiyama-Okurari-ryō no jinkō hendō to sono chiikiteki tokuchō* (Population Change and Its Special Regional Features in the Aizu/Minamiyama-Okurari-ryō Region during the Eighteenth and Nineteenth Centuries). Unpublished manuscript.
Kano, H.
1974   "Fatigue in Man, Skill and Systems: Recollection of Fifty Years of Psychological Approaches in ISL." *Journal of Science of Labour* 50, no. 12 (Pt. 2): 871–891.
Katz, S. H.
1987   "Fava Bean Consumption: A Case for the Co-Evolution of Genes and Culture." In *Food and Evolution: Toward a Theory of Human Food Habits,* ed. M. Harris and E. B. Ross, 133–159. Philadelphia: Temple University Press.
Kikuchi, T.
1986   *Shinden kaihatsu* (Development of New Rice Fields). Tokyo: Kokoninsha.
Kito, H.
1983   *Nihon nisennen no jinkōshi* (Two Thousand Years of Japanese Population History). Tokyo: PHP Kenkyūshō.
1986   "Meiji senki no shushoku kōsei to sono chiiki pataan" (The Composition and Regional Pattern of the Major Type of Food Consumption in Early Meiji). *Jochi Keizai Ronshū* 31 (2): 30–43.

Klamer, A., D. N. McCloskey, and R. M. Solow, eds.

1988   *The Consequences of Economic Rhetoric.* Cambridge: Cambridge University Press.

Komlos, J., ed.

1994   *Stature, Living Standards, and Economic Development: Essays in Anthropometric History.* Chicago: University of Chicago Press.

Kurosaki, C.

1967   "Shushoku shōhi no chiikiteki keikō" (Regional Trends in Food Consumption for the Major Foods Consumed). *Waseda Daigaku Kōtōgakuin Kenkyū Nenshi 12 Go: 4–26.*

1983   "Monmōritsu teika no chiikiteki dōkō" (Regional Trends in the Decline of Illiteracy). *Rekishi Chirigaku Kihyō 25: 21–41.*

Kuznets, S.

1971   *Economic Growth of Nations: Total Output and Production Structure.* Cambridge: Harvard University Press.

Levine, S. B., and H. Kawada

1980   *Human Resources in Japanese Industrial Development.* Princeton: Princeton University Press.

Maddala, G. S.

1992   *Introduction to Econometrics,* 2d ed. New York: Macmillan.

McCloskey, D. N.

1994   *Knowledge and Persuasion in Economics.* Cambridge: Cambridge University Press.

Miller, R. A.

1982   *Japan's Modern Myth: The Language and Beyond.* New York: Weatherhill.

Mirowski, P.

1987   "Shall I Compare Thee to a Minkowski-Ricardo-Leontief-Metzler Matrix of the Mosak-Hicks Type? Or, Rhetoric, Mathematics and the Nature of Neoclassical Economic Theory." *Economics and Philosophy 3: 67–96.*

Miura, T.

1978   *Rōdō to kenkō no rekishi, daiichikan—kodai kara Bakumatsu made* (The History of Labor and Health, Vol. 1: From Ancient Times until the Bakumatsu Period). Kawasaki: Rōdōkaakukenkyūjō.

1980a   *Rōdō to kenkō no rekishi, dainikan—Meiji shonen kara kōbahō jicchi made* (The History of Labor and Health, Vol. 2: From the Beginning of Meiji until the Implementation of the Factory Law). Kawasaki: Rōdōkagakukenkyujō.

1980b   *Rōdō to kenkō no rekishi, daisankan—Ohara rōdōkagaku-kenkyūjō no sōritsu kara Showa e* (The History of Labor and Health, Vol. 3: From the Founding of the Ohara Research Institute for the Science of Labor until the Showa Period). Kawasaki: Rōdōkagakukenkyūjō.

1981   *Rōdō to kenkō no rekishi, daiyonkan—jugonensensōka no rōdō to kenkō* (The History of Labor and Health, Vol. 4: Labor and Health under Conditions of Fifteen Years of War). Kawasaki: Rōdōkaakukenkyūjō.

1986   "A Short History of Occupational Health in Japan (Pt. 2): From the Early Meiji Period to the End of the Taisho Period (1886–1926)." *Journal of Science of Labour* 62, no. 2 (Pt. 2): 1–17.

Morris, D., and T. C. Smith

1985   "Fertility and Mortality in an Outcaste Village, 1750–1869." In *Family and Population in East Asian History*, ed. S. B. Hanley and A. P. Wolfe, 229–246. Stanford: Stanford University Press.

Mosk, C.

1983   *Patriarchy and Fertility: Japan and Sweden, 1880–1960.* New York: Academic Press.

1995a   "Household Structure and the Labor Market in Prewar Japan." *Journal of Family History* 20, no. 1: 103–125.

1995b   "Small-Scale Production and Urban Expansion in Industrializing Japan: Nagoya, 1890–1940." Paper presented to the Roundtable "Urban Demography During Industrialization," at the XVIIIᵗʰ Congress of Historical Studies, Montreal, 30 August.

1995c   *Competition and Cooperation in Japanese Labour Markets.* Houndmills, Basingstoke, Hampshire: Macmillan.

Mosk, C., and S. Pak

1978   "Food Consumption, Physical Characteristics and Population Growth in Japan, 1874–1940." Department of Economics, University of California at Berkeley, Working Paper no. 102.

Nakane, C.

1967   *Kinship and Economic Organization in Rural Japan.* New York: Humanities Press.

Nelson, J. S., A. Megill, and D. N. McCloskey

1987   *The Rhetoric of the Human Sciences: Language and Argument in Scholarship and Public Affairs.* Madison: University of Wisconsin Press.

Nihon Koseihō (Japan Ministry of Health)

1955   *Isei 80 nenshi* (Eighty-Year History of Medical Policy). Tokyo: Insatsukyoku Choyosha.

Nihon Naikaku Tōkeikyoku (Japan Cabinet Bureau of Statistics)

various   *Nihon teikoku tōkei nenkan* (Statistical Yearbook of the Empire
years   of Japan). Tokyo: Naikaku Tōkeikyoku.

1932   *Nōgyō chōsa kekka* (Report of the Results of the Agricultural Census). Tokyo: Naikaku Tōkeikyoku.

Nihon Nōrinshō Nōmukyoku (Japan Ministry of Agriculture and Forestry Agricultural Bureau)

1953   *Nōka keizai chōsa hōkoku* (Report of the Surveys on Economy of Agricultural Families). Tokyo: Nōrintōkeikai.

Nishida, Y.
1986   "Growth of the Meiji Landlord System and Tenancy Disputes
          after World War I: A Critique of Richard Smethurst, *Agricultural
          Development and Tenancy Disputes in Japan, 1870–1940.*"
          Journal of Japanese Studies 15 (2): 389–415.
Ohkawa, K., and M. Shinohara
1979   *Patterns of Japanese Economic Development: A Quantitative
          Appraisal.* New Haven: Yale University Press.
Ohta, M., and M. Sawayama
1994   "An Analysis of the Motivation for *Mabiki* and Abortion as Re-
          lated to Child Rearing Customs in Early Modern Japan: Re-
          search Based on Non-Quantitative Evidence of the History of
          the People's Mentality." Paper presented to the Workshop on
          Abortion, Infanticide and Neglect in the Asian Past: Japan in
          East Asian Perspective, 20–21 October, International Research
          Center for Japanese Studies, Kyoto.
Onishi, Y.
1988   "Wagakuni no fukushi kokka no tenkai" (The Evolution of
          Japan's Welfare Nationalization). In *Fukoshi kokka no rek-
          ishi to tenbo,* ed. M. Adachi, 81–110. Kyoto: Horitsub-
          unkasha.
Pellett, P. L.
1987   "Problems and Pitfalls in the Assessment of Human Nutritional
          Status." In *Food and Evolution: Toward a Theory of Human
          Food Habits,* ed. M. Harris and E. B. Ross, 163–180. Philadel-
          phia: Temple University Press.
Pike, R. L., and M. L. Brown
1967   *Nutrition: an Integrated Approach.* New York: John Wiley and
          Sons.
Pindyck, R. S., and D. L. Rubinfeld
1991   *Econometric Models and Economic Forecasts.* New York: Mc-
          Graw-Hill.
Ramseyer, J. M.
1989   "Water Law in Imperial Japan: Public Goods, Private Claims,
          and Legal Convergence." *Journal of Legal Studies* 18 (1):
          51–77.
Saito, O.
1991   "Nōgyō hatten to josei rōdō: Nihon no rekishiteki keiken"
          (Women's Labor in Agricultural Development: Japan's Experi-
          ence). *Keizai Kenkyō* 42: 31–41.
1993   "Infant Mortality in Pre-Transition Japan: Levels and Trends."
          Institute of Economic Research, Hitotsubashi University, Dis-
          cussion Paper no. 273.
Sanderson, F. H.
1978   *Japan's Food Prospects and Policies.* Washington, D.C: Brook-
          ings Institution.

Sen, A.
    1987   "Poverty and Entitlements." In *Food Policy,* ed. J. P. Gittinger,
              J. Leslie, and C. Hoisington, 198–204. Baltimore: Johns Hop-
              kins University Press.
    1990   "Food, Economics and Entitlements." In *The Political Economy
              of Hunger,* Vol. 1: *Entitlement and Well-Being,* ed. J. Dreze and
              A. Sen, 34–52. Oxford: Clarendon Press.
Sen, A., J. Muelbauer, R. Kanbur, K. Hart, and B. Williams [edited by Geoffry
Hawthorn]
    1987   *The Standard of Living: The Tanner Lectures, Clare Hall, Cam-
              bridge, 1985.* Cambridge: Cambridge University Press.
Shay, T.
    1994   "The Level of Living in Japan, 1885–1938: New Evidence." In
              *Stature, Living Standards, and Economic Development: Essays
              in Anthropometric History,* ed. J. Komlos, 173–201. Chicago:
              University of Chicago Press.
Shinohara, M.
    1965   *Chiiki keizai kōzō no keiryōteki bunseki* (Quantitative Analysis
              of Regional Economic Structure). Tokyo: Iwanami Shoten.
Smethurst, R.
    1986   *Agricultural Development and Tenancy Disputes in Japan,
              1870–1940.* Princeton: Princeton University Press.
Smith, R.
    1983   "Making Village Women into 'Good Wives and Wise Mothers'
              in Prewar Japan." *Journal of Family History* 8 (1): 70–84.
Smith, T.
    1959   *The Agrarian Origins of Modern Japan.* Stanford: Stanford Uni-
              versity Press.
    1988   *Native Sources of Japanese Industrialization.* Berkeley: Univer-
              sity of California Press.
Steckel, R.
    1994a   "Heights and Health in the United States, 1710–1950." In
              *Stature, Living Standards, and Economic Development: Essays
              in Anthropometric History,* ed. J. Komlos, 153–170. Chicago:
              University of Chicago Press.
    1994b   "Stature and the Standard of Living." Paper presented to the
              1994 Annual Meetings of the Social Science History Associa-
              tion, October, Atlanta, Georgia.
Sugaya, A.
    1976   *Nihon iryō seidoshi* (A History of Japan's Medical System).
              Tokyo: Harashobo.
    1982   *Nihon iryō seisakushi* (A History of Japanese Medical Policy).
              Tokyo: Nihon Hyōronsha.
Sugita, G.
    1969   *Dawn of Western Science in Japan* [Translation of *Rangaku Ko-
              tohajime,* by R. Matsumoto]. Tokyo: Hokuseido Press.

Tanner, J. M.
1961   *Education and Physical Growth: Implications of the Study of Children's Growth for Educational Theory and Practice.* London: University of London Press.
1978   *Foetus into Man: Physical Growth from Conception to Maturity.* Cambridge: Harvard University Press.
1981   *A History of the Study of Human Growth.* Cambridge: Cambridge University Press.
1994   "Introduction: Growth in Height as a Mirror of the Standard of Living." In *Stature, Living Standards, and Economic Development: An Anthropometric History,* ed. J. Komlos, 1–6. Chicago: University of Chicago Press.

Taylor, T. G.
1982   *Nutrition and Health.* London: Edward Arnold.

Teruoka, G.
1942   "Kokuminshoku ni okeru tanpaku hitsujuryō ni tsuite no kihonteki chōsa hōkoku" (Report of the Basic Survey Concerning the Demand for Proteins in the National Diet). *Rōdō Kagaku* 19: 383–495.

Tilly, L.
1983   "Food Entitlement, Famine and Conflict." *Journal of Interdisciplinary History* Vol. 14(2): 205–226.

Tomobe, K.
1994   "Coale-Trussel Indices, Breast Feeding, and Infanticide in Tokugawa Japan." Paper presented to the Workshop on Abortion, Infanticide and Neglect in the Asian Past: Japan in East Asian Perspective, 20–21 October, International Research Center for Japanese Studies, Kyoto.

Totman, C. D.
1967   *Politics in the Tokugawa Bakufu: 1600–1843.* Cambridge: Harvard University Press.

Tsuchiya, K.
1976   *Productivity and Technological Progress in Japanese Agriculture.* Tokyo: University of Tokyo Press.

Tsurumi, E. P.
1990   *Factory Girls: Women in the Thread Mills of Meiji Japan.* Princeton: Princeton University Press.

Uchino, S.
1977   *Jinkō idō to shoku seikatsu—Toshika ni tomo na jinkō idō to shokuji naiyō no henka* (Population Mobility and Food Life—Population Movements Accompanying Urbanization and Changes in the Content of Diet). Tokyo: Daiichi Shuppan Kabushiki Gaisha.

Umemura, M., K. Akasaka, R. Minami, N. Takamatsu, K. Arai, and S. Itoh
1988   *Manpower.* Tokyo: Toyo Keizai Shinposha. Volume 2 of *Long-Term Historical Statistics of Japan Since 1868.*

Umemura, M., N. Takamatsu, and S. Itoh
    1983   *Regional Economic Statistics.* Tokyo: Toyo Keizai Shinposha.
                Volume 13 of *Long-Term Historical Statistics of Japan Since
                1868.*
Umemura, M., S. Yamada, Y. Hayami, N. Takamatsu, and M. Kumazaki
    1966   *Agriculture and Forestry.* Tokyo: Toyo Keizai Shinposha.
Vlastos, S.
    1986   *Peasant Protests and Uprisings in Tokugawa Japan.* Berkeley,
                Los Angeles, and London: University of California Press.
Waswo, A.
    1977   *Japanese Landlords.* Berkeley, Los Angeles, and London: Uni-
                versity of California Press.
    1982   "In Search of Equity: Japanese Tenant Unions in the 1920s." In
                Conflict in Modern Japanese History: The Neglected Tradition,
                ed. T. Najita and V. Koschmann, 366–411. Princeton: Princeton
                University Press.
Wallace, B.
    1972a   *Genetics, Evolution, Race, Radiation Biology: Essays in Social
                Biology.* Vol. 2. Englewood Cliffs, N.J.: Prentice-Hall.
    1972b   *Disease, Sex, Communication, Behavior: Essays in Social Biol-
                ogy.* Englewood Cliffs, N.J.: Prentice-Hall.
Yagi, T.
    1935   "Rōdōsha saiyoji no mitai kensahō: Toku ni sono keitaigaku
                keisoku hōmen ni tsuite" (Methods of Physical Examination of
                Freshly Inducted Workers: Especially Viewed from the Vantage
                of the Science of Measurement of Physique). *Rōdoō Kagaku
                Kenkyū* 11: 629–684.
Yamamura, K.
    1974   *A Study of Samurai Income and Entrepreneurship: Quantitative
                Analyses of Economic and Social Aspects of the Samurai in
                Tokugawa and Meiji Japan.* Cambridge: Harvard University
                Press.
Yamanaka, H.
    1966   *Seimei hoken kinyū hattenshi* (A History of the Financial Devel-
                opment of Life Insurance). Tokyo: Yuhikaku.
Yasuba, Y.
    1986   "Bakumatsu no seikatsu suijun ni tsuite" (A Note on the Stan-
                dard of Living in the Bakumatsu Period). *Osaka Economic Pa-
                pers* 35 (4): 126–130.
    1987   "The Tokugawa Legacy: A Survey." *Economic Studies Quar-
                terly* 38 (4): 290–308.

Index

Designer: UC Press Staff
Compositor: Impressions Book and Journal Services, Inc.
Text: Sabon
Display: Sabon
Printer: Edwards Brothers, Inc.
Binder: Edwards Brothers, Inc.